T0140771

Performance and Thermal Management on Self-adaptive Hybrid Multi-cores

Dissertation

A thesis submitted to the
Faculty of Computer Science, Electrical Engineering, and Mathematics
of the
University of Paderborn
in partial fulfillment of the requirements for the
degree of *Dr. rer. nat.*

by

Markus Happe

Paderborn, Germany
Date of submission: 27.02.2013

Bibliografische Information der Deutschen Nationalbibliothek

Die Deutsche Nationalbibliothek verzeichnet diese Publikation in der
Deutschen Nationalbibliografie; detaillierte bibliografische Daten sind
im Internet über http://dnb.d-nb.de abrufbar.

©Copyright Logos Verlag Berlin GmbH 2013
Alle Rechte vorbehalten.

ISBN 978-3-8325-3425-7

Logos Verlag Berlin GmbH
Comeniushof, Gubener Str. 47,
10243 Berlin
Tel.: +49 (0)30 42 85 10 90
Fax: +49 (0)30 42 85 10 92
INTERNET: http://www.logos-verlag.de

*" I almost wish I hadn't
gone down that rabbit-hole
—and yet—and yet—
it's rather curios, you know,
this sort of life! "*

Alice, *Alice's Adventures In Wonderland*

Supervisor:
   Prof. Dr. Marco Platzner

Reviewers:
   Prof. Dr. Marco Platzner
   Prof. Dr. David Andrews
   Prof. Dr. Franz J. Rammig

Additional members of the oral examination committee:
   Prof. Dr. Ulrich Rückert
   Dr. Matthias Fischer

Date of submission:
   February 27, 2013

Date of public examination:
   April 19, 2013

Abstract

Over the last years system-on-chip architectures evolved from single-processor systems over homogeneous multi-processor systems to heterogeneous multi-processor systems in order to keep up with the ever growing demand for computational efficiency. Furthermore, more and more (reconfigurable) hardware accelerators have found their way into these computing systems, because these accelerators can provide substantial performance speedups and/or power consumption reductions for certain task types. Several research projects have investigated new hybrid computer architectures that combine heterogeneous processors with reconfigurable hardware accelerators on a single chip. Recently, programming models and operating systems have been developed to ease the programming and execution of applications for these architectures.

However, as the computing devices increase in size and allow for large and complex multi-core architectures, there are further—yet unsolved—challenges besides the programming and execution of applications on such platforms. Due to unforeseeable system dynamics—such as varying user performance requirements, changing task workloads and dynamic task sets—many system characteristics such as the application's performance, on-chip temperature distribution and overall power consumption have to be observed at run-time. The run-time management of system resources and physical aspects—like temperature and power consumption—-imposes a great challenge, which has been hardly investigated for hybrid multi-core systems so far.

In this thesis, we propose novel run-time management techniques for an emerging architecture, a self-adaptive hybrid multi-core. The hybrid multi-core contains heterogeneous processors that execute software threads and reconfigurable hardware accelerators that execute so-called hardware threads. Using internal local sensors and dedicated system models, such multi-core systems are able to capture and analyze the system state at run-time, and adapt the thread-to-core mapping in order to meet user-defined goals. Due to the dissimilar nature of processors and hardware cores, different thread implementations show varying performance, power and thermal characteristics. These heterogeneous characteristics introduce novel challenges, which can not be met with traditional run-time management techniques that target homogeneous architectures. This work focuses on *autonomous performance and thermal management* techniques.

For *performance management*, novel models and algorithms are discussed that target streaming applications, which are a typical workload for embedded systems. The proposed self-adaptive techniques can dynamically add and remove thread instances in order to meet user-defined performance constraints such as a lower bound while minimizing the number of active cores at the same time. To demonstrate the effectiveness of the self-adaptation techniques, a detailed experimental evaluation of a real-world case study, which has been implemented on an FPGA, is presented.

For *thermal management*, this thesis introduces a monitoring system with which the hybrid multi-core system can capture core-specific temperature information. This work presents a detailed study of ring oscillator-based sensor layouts along with a novel self-calibration technique of such sensors, which eliminates the need of an extensive manual calibration using external devices. We propose a novel approach with which the system can autonomously learn the parameters of a state-of-the-art thermal model at run-time. The thermal model can then be used by the system to predict possible effects of thread remappings to the on-chip temperature distribution. Finally, this work presents and evaluates several thermal management techniques for self-adaptive hybrid multi-cores, where the strategies differ in the degree of knowledge about the system.

Zusammenfassung

In den letzten Jahren haben sich „System-on-Chip"-Architekturen von Ein-Prozessor-Systemen über homogene Mehr-Prozesor-Systeme bis hin zu heterogenen Mehr-Prozessor-Systemen weiterentwickelt, um die stetig ansteigenden Anforderungen an die Rechenleistung gerecht zu werden. Außerdem finden (rekonfigurierbare) Hardware-Beschleuniger vermehrt Einsatz in „System-on-Chip"-Architekturen, da diese Beschleuniger für gewisse Applikationsprozesse zu einer erheblichen Beschleunigung der Ausführungszeiten und/oder einer deutlichen Reduzierung der Leistungsaufnahme betragen können. Verschiedene Forschungsprojekte haben daher neue eingebettete Rechnerarchitekturen erforscht, welche heterogene Prozessoren mit rekonfigurierbaren Hardware-Beschleunigern auf einem einzelnen Chip vereinen. Seit wenigen Jahren werden Programmiermodelle und Betriebssysteme entwickelt, welche die Entwicklung und Ausführung von Applikationen für diese Architekturen vereinfachen sollen.

Da die zugrunde liegenden Chips immer größer werden und daher immer komplexere Systeme gebaut werden können, gibt es neben der Programmierbarkeit und der Ausführung von Applikationen noch weitere wichtige Herausforderungen, welche bis jetzt ungelöst sind. Durch unvorhersehbare Systemdynamiken, wie zum Beispiel wechselnden Anforderungen an die Ausführungszeiten von Applikationen, dynamischen Arbeitslasten oder variierende Prozessmengen, müssen bestimmte Systemeigenschaften wie die Applikationslaufzeiten, die Temperaturverteilung auf dem Chip sowie die Gesamt-Leistungsaufnahme des Systems zur Laufzeit verfolgt werden. Das Laufzeitmanagement der Systemressourcen oder der physikalischen Aspekte wie zum Beispiel der Temperaturverteilung oder der Leistungsaufnahme wirft große Herausforderungen auf.

In der vorliegenden Arbeit stellen wir neuartige Techniken für das Laufzeitmanagement vor, welche auf eine neu entstehende Architektur, einem selbst-adaptiven hybriden Mehrkern, abzielen. Dieser hybride Mehrkern besteht einerseits aus verschiedenen heterogenen Prozessoren, welche Software-Threads ausführen, und anderseits aus rekonfgurierbaren Hardware-Beschleunigern, welche sogenannten „Hardware-Threads" ausführen. Mittels lokaler interner Sensoren und dedizierten Systemmodellen können diese selbst-adaptiven hybriden Mehrkern-Systeme zur Laufzeit den Systemzustand bestimmen und analysieren sowie die aktuelle Abbildung von Threads auf Rechenkerne anhand der benutzerdefinierten Zielen anpassen. Durch die unterschiedliche Beschaffenheiten von Prozessoren und Hardware-Beschleunigern, können verschiedene Implementierung ein und desselben Threads unterschiedliche Laufzeiten und Leistungsaufnahmen haben sowie verschiedene Temperaturen generieren. Diese unterschiedlichen Eigenschaften stellen neue Herausforderungen auf, welche mit herkömmlichen Laufzeitmanagementverfahren, die ihrerseits für homogene Systeme entwickelt wurden, nicht gelöst werden können. Diese Arbeit fokussiert dabei auf das *Laufzeitmanagement zur effizienten Ausführung von Applikationen* sowie auf das *thermische Management des Chips*.

Für das *Laufzeitmanagement zur effizienten Ausführung von Applikationen* werden neuartige Modelle und Algorithmen vorgestellt, die sich auf Streaming-Anwendungen spezialisieren, was einen typischen Anwendungsfall für eingebettete Systeme darstellt. Die vorgestellten selbst-adaptiven Techniken können Thread-Instanzen dynamisch zur Laufzeit zu dem System hinzufügen oder von diesem wieder entfernen, um dadurch auf der einen Seite die benutzerdefinierte Anforderungen an die Ausführungszeit der Applikation zu erfüllen und auf der anderen Seite zur gleichen Zeit die Anzahl der verwendeten Rechenkerne zu minimieren. Um die Effizienz der selbst-adaptiven Techniken zu demonstrieren, werden die Techniken anhand einer realistischen Fallstudie, welche auf einem FPGA implementiert wurde, umfangreich untersucht.

Für das *thermische Management* führt die vorliegende Arbeit ein Überwachungssystem ein, mit welchem das hybride Mehrkern-System Temperaturinformationen zu den einzelnen Kernen abrufen kann. Diese Arbeit präsentiert dazu eine detaillierte Studie über Sensor-Layouts, welche auf Ring-Oszillatoren basieren, sowie einer neuartigen Selbst-Kalibrierungstechnik für solche Sensoren, welche die Notwendigkeit einer umfangreichen, manuellen Kalibrierung mittels externer Geräte beseitigt. Wir schlagen zudem einen neuartigen Ansatz vor, mit welchem das hybride Mehrkern-System die Parameter seines thermischen Modells, welches dem aktuellen Stand der Wissenschaft entspricht, autonom zur Laufzeit lernt, um dadurch die möglichen Auswirkungen einer Selbst-Adaption auf Thread-Ebene vorhersagen zu können. Schließlich stellt diese Arbeit mehrere Techniken für das thermische Laufzeitmanagement für selbst-adaptive, hybride Mehrkern-Systeme vor, welche sich anhand ihres Vorwissens unterscheiden, und bewertet diese.

Contents

The White Rabbit put on his spectacles. 'Where shall I begin, please your Majesty?' he asked.
'Begin at the beginning,' the King said gravely, 'and go on till you come to the end: then stop.'
These were the verses the White Rabbit read:

Lewis Carroll, *Alice's Adventures In Wonderland*

CHAPTER 1

Introduction

Three trends emerged in the last decade for system-on-chips: the use of (i) multiple processors instead of a single processor, (ii) the use of heterogeneous processors instead of homogeneous processors and (iii) the use of additional hardware accelerators such as FPGA[1]-based reconfigurable co-processors. Single processor systems have evolved to multi-processor systems because the clock frequency of single processors has hit the so-called power wall. The power wall occurs due to the current trend that a gradual increase in the processor's clock frequency leads to a polynomial increase of power consumption. This drastic raise in power consumption is caused by a linear increase of the dynamic power due to the increased frequency, on the one side, and several indirect factors, on the other side. For instance, a higher input voltage is required to drive the raised frequency of the processor, which effects the power consumption. To allow high frequencies, transistors have to be redesigned, which typically results in a higher leakage current. When the aforementioned effects lead to rising temperatures, this in turn raises the resistances and, therefore, further increases the power dissipation.

For many years, the growing number of transistors per die could be used to improve the processor's architecture for performance-efficiency, for instance by optimizing the branch prediction, the pipelining, and, the super-scalar and out-of-order execution. However, the development of single core processors has reached its limits because the costs for further improvements of the processor architecture started to outweigh the benefits. To further increase the performance

---

[1]field-programmable gate array (FPGA)

of computing systems, multiple processors are placed on the same die nowadays. Multiple applications or application tasks can be executed in parallel to reach a further speedup in the system's overall performance.

Heterogeneous multi-processor systems can be superior to homogeneous multi-processor systems if the software tasks can make use of the heterogeneity. For instance, a software task can achieve a higher performance in execution time when it is mapped to the processor that shows the greatest affinity towards that task. Furthermore, the processors can differ in the clock frequency, which does not only influence the performance but also the power consumption and thermal characteristic of that core. Thus, task migration between heterogeneous processors can be also used to control the power consumption and the temperature profile of the chip.

Hybrid multi-core systems combine heterogeneous multi-processor systems with reconfigurable hardware, which hosts additional cores. Reconfigurable hardware cores show great potential in high performance, low power consumption and, therefore, reduced thermal leakage for certain classes of tasks. Usually, tasks with a high degree of parallelism are well-suited for a hardware implementation. In contrast, control-dominated tasks show limited potential for parallelism and should be mapped to software. The application designer is in charge to define the hardware/software partitioning due to missing industrial tool support. Moreover, a widely-used common programming model and execution environment is missing. This forces application designers to manually integrate their reconfigurable hardware cores in their software applications.

Several research projects, such as ReconOS [96] and HThreads [113], have extended the widely-used multithreading approach to reconfigurable hardware to overcome this burden. Using the multithreading approach as programming and execution model eases the development of hybrid multi-core systems since the thread designer can focus on his specific thread implementation. In the multithreading paradigm, threads share common resources such as memory, communication primitives and synchronization primitives. In ReconOS, for instance, hardware and software threads can access the same operating system resources (mutexes, semaphores, message boxes, etc.) using the well-known POSIX[2] API[3]. This eases the programming of such complex systems.

As the computing devices increase in size and allow for large and complex multi-core architectures, there are further—yet unsolved—challenges besides the programming and execution of corresponding applications. Due to unforeseeable system dynamics—such as varying user performance requirements, changing task workloads and dynamic task sets—many system characteristics such as the

---

[2]portable operating system interface (POSIX)
[3]application programmer interface (API)

application's performance, on-chip temperature distribution and overall power consumption have to be observed at run-time. The run-time management of the system resources and physical aspects like temperature and power consumption imposes a great challenge on designers of hybrid multi-core systems. This thesis investigates the performance and temperature management on such systems using self-adaptation techniques at thread-level to partly solve this challenge.

1.1 Claim of This Thesis

As target platform for future embedded systems, we propose a self-adaptive hybrid multi-core that consists of heterogeneous processors—executing software threads— and reconfigurable hardware cores—executing so-called hardware threads. Both hardware and software threads access shared memory and can call the same operating system services. In this thesis we will use ReconOS [95]—that has successfully extended the multithreading approach to reconfigurable hardware— as a basis to build self-adaptive hybrid multi-cores.

To enable self-adaptation, we propose to use internal monitors like timers (to measure the execution time of threads) and temperature sensors (to measure the on-chip temperature distribution). The signals of the internal monitors serve as input for the system's goal-oriented models that are either used to trigger self-adaptation in a reactive manner or to train & correct the model parameters of more advanced self-adaptation strategies.

For self-adaptation, we propose to use thread migration between different cores of the hybrid multi-core. Due to the different nature of processors and hardware cores, different thread implementations show varying performance, power and thermal characteristics. Thus, migrating a thread from a processor to a hardware core can result in an improved performance for this thread. However, the number of hardware cores in the system is limited. Hence, the system might need to adapt the thread-to-core mappings depending on the current workload of the active threads.

Furthermore, it is not always possible to preempt a hardware or a software thread and transfer its context to a core with a different modality (hardware → software, software → hardware). Thus, we use cooperative multitasking, which allows us to migrate threads between the hardware/software boundary on suitable migration points, where the thread context is well-defined and minimal. For self-adaptation it is beneficial to minimize the number of thread migrations, especially if hardware threads are involved. Migrating a thread to a hardware core requires a (partial) reconfiguration of the core, which results in a high overhead in execution time.

If the system has the goal to avoid local temperature hot spots, the system could monitor the current thermal profile of the chip using internal temperature sensors. To balance the on-chip temperature, threads from 'hot' cores can be migrated into cold regions in a reactive manner. A reactive migration mechanism would possibly lead to a high number of thread migrations if the underlying thermal model of the chip is unknown. Hence, advanced thermal models would be beneficial that can predict the thermal effects of thread migrations and, thus, select the optimal thread mapping. In this way, the number of thermally-triggered costly thread migrations can be reduced.

Predicting the thermal effects of a thread migration on a hybrid multi-core is challenging, because the thermal characteristic of a thread mapped to a core depends on several criteria, such as the type of thread implementation (software thread, hardware thread), the current workload, the heat transfer of neighboring cores—where the cores can have different spatial shapes on the chip—and even the process variations of the chip regions.

Despite the mentioned challenges, this thesis formulates the following claim.

> **Claim:** Hybrid multi-cores can effectively perform performance and thermal management autonomously at run-time using self-adaptation on thread-level.

1.2 Contributions of This Thesis

In this thesis, we propose a self-adaptive hybrid multi-core architecture that consists of (heterogeneous) processors, reconfigurable hardware cores and monitoring units (to capture the system's state). Furthermore, this thesis contributes novel self-adaptation models and self-adaptation techniques that are applied for autonomous performance and thermal management.

In particular, we provide the following contributions:

- We have developed an FPGA-based self-adaptive hybrid multi-core prototype based on the multithreading programming and execution environment ReconOS. The prototype contains internal sensors to capture the system's state. As performance monitors, we use timers to measure the execution time of threads. Furthermore, we have developed novel self-calibrating temperature sensors, which can measure the on-chip temperature distribution. Similar to related work, we use ring oscillators as temperature sensors. We employ dedicated regional heat-generating cores to create global heat and calibrate our sensors according to a pre-calibrated built-in thermal diode. Thus, an extensive manual calibration of the sensors using external devices—such as temperature controlled ovens or infrared cameras—can be avoided. Finally, our prototype emulates thread migration between (hybrid) cores using dynamic thread creations and terminations.

- We have contributed to the further development of the programming and execution model ReconOS and its corresponding tool flow to state-of-the-art FPGA families. Significant changes in the FPGA fabric, the supported communication buses, and the tool flow required a drastic reconstruction of the ReconOS system architecture and its tool flow. For the new ReconOS version, we have implemented operating system calls as part of the operating system interface of hardware threads and burst transfers to shared memory as part of the memory interface of hardware threads. Furthermore, we have generated new reference designs, e.g., a design with cache support for the main processor, and ported a video object tracking case study to the new ReconOS version.

- For autonomous performance management on self-adaptive hybrid multi-cores, we have developed performance models and self-adaptation algorithms that aim to meet user-defined performance constraints. Our self-adaptation techniques either focus on guaranteeing a lower bound for a user-defined application performance (if possible) or seek to keep the application performance within a user-defined interval. In this thesis, we define parallel applications as target applications for an autonomous performance management. The self-adaptation techniques take advantage of the different affinities of a thread to specific cores and migrate threads between cores with the goal to minimize the number of active cores, which can result in a reduced overall power consumption.

  For evaluation of our autonomous performance management techniques, we have applied our novel strategies on a real-world case study. Our case study is a particle filter-based video object tracker. All threads have been implemented as software and hardware threads and can be migrated between cores at run-time. Thus, we were able to demonstrate the applicability of the proposed self-adaptation techniques to a realistic application scenario.

- For autonomous temperature management, we have created a novel thermal model that can predict the thermal effects of thread migrations on the temperature distribution. Like related work, i.e. HotSpot, we use resistor/capacitor networks to model the heat flow. Unlike HotSpot, we are mainly interested in the temperature distribution inside the silicon layer and therefore, we have reduced the granularity of the thermal model.

Furthermore, the static parameters of the thermal model are learned in an initial learning phase of the system-on-chip (SoC). Active threads that are mapped onto specific cores are modeled as (dynamic) heat sources. These heat-sources are thread-specific and are learned at run-time. Our self-adaptation techniques aim to balance the on-chip temperature distribution to avoid local hot spots. Therefore, the algorithms migrate threads between the cores to balance the chip temperature. Because the system can continuously learn its thermal model, the migration decisions may improve over time.

1.3 Thesis Outline

The remainder of this thesis is structured as follows.

Chapter 2 provides background and related work on hybrid and/or self-adaptive multi-cores. A special focus is set on performance and thermal management on multi-core systems.

Chapter 3 outlines concepts and key ideas for self-adaptive hybrid multi-cores based on multithreading. The chapter introduces thread migration as self-adaptation technique to fulfill different optimization goals. The ReconOS multithreading programming model is therefore extended from hardware threads towards hybrid hardware/software threads. Finally, the chapter formalizes the goals for performance and thermal management.

Chapter 4 presents the proposed self-adaptive hybrid multi-core architecture, the operating system layer and the corresponding tool flow. The hybrid multi-core builds on the ReconOS execution environment [96]. The chapter highlights the extensions that were introduced to support self-adaptation— such as monitoring cores—and to support state-of-the-art FPGA families.

Chapter 5 introduces the performance model and the performance-driven self-adaptation mechanisms. The multi-core self-adapts as a result of a trade-off between the number of active cores—that currently execute threads—and the application's performance, which has to meet user-defined quality-of-service constraints. As examples for typical quality-of-service constraints, the overall application performance (i) has to stay above a user-defined performance bound or (ii) to stay inside a user-defined performance interval. To evaluate the applicability of the self-adaptation methods, they are applied to a real-world case study. Therefore, the chapter presents a particle filter-based video object tracker as real world case study for later experimental evaluation.

Chapter 6 defines the thermal model of an FPGA-based hybrid multi-core and introduces methods for measuring the on-chip temperature distribution and dedicated heat-generating cores to create spatial temperature differences. The thermal model is more advanced than the performance model in Chapter 5. This chapter shows how the model parameters can be learned at run-time to predict future on-chip temperature distributions and the thermal effects of thread migration. Furthermore, thermally-driven self-adaptation techniques are presented and discussed. Finally, a temperature simulator is presented that is used to evaluate the self-adaptation techniques.

Chapter 7 experimentally evaluates the proposed performance and thermal models and the corresponding self-adaptations techniques. To determine the efficiency of the self-adaptation techniques for performance management, the techniques are experimentally applied to a real-world case study, a video object tracker, on a Xilinx Virtex-4 FPGA.

For thermal management, the chapter presents the results on temperature measurements using a self-calibrated sensor grid and provides experimental proof that high spatial temperature differences can already be generated on today's FPGA systems. Furthermore, the chapter presents experimental results for learning the thermal model of an FPGA at run-time on a Virtex-6 FPGA. The self-adaptation techniques for thermal management are evaluated using simulations.

Chapter 8 makes the case for integrating self-awareness into hybrid multi-cores. The chapter describes the concepts of self-awareness and presents a blueprint for future self-aware hybrid multi-cores.

Chapter 9 summarizes the contributions and results of this thesis. Finally, the chapter gives an outlook to possible future research directions of self-adaptive/self-aware hybrid multi-cores.

For, you see, so many out-of-the-way things had happened lately, that Alice had begun to think that very few things indeed were really impossible.

Lewis Carroll, *Alice's Adventures In Wonderland*

CHAPTER 2

Background and Related Work

This chapter provides background on field-programmable gate arrays and presents related work on hybrid multi-core systems, approaches for task migration across the hardware/software boundary, and, finally, performance and thermal management techniques on multi-core systems.

2.1 Field-programmable Gate Arrays

Field-programmable gate arrays (FPGAs) are integrated circuits where the logic blocks and the interconnect are configurable. The first FPGAs were developed by the company Xilinx in the 1980s. Nowadays the companies Xilinx and Altera are the two most prominent commercial vendors of FPGAs [99].

FPGAs contain a huge array of configurable logic blocks (CLBs). These CLBs contain a number of programmable look-up tables (LUTs) and flip-flops (FFs). The LUTs have multiple inputs and up to two outputs. An arbitrary function can be configured into an LUT, which translates the input signals to an output signal. The output values can be stored in flip-flops, which are directly connected to LUTs. Flip-flops are synchronous storage elements, which can change their values at each clock cycle. Furthermore, FFs can also be used asynchronously as latches. A large number of CLBs can be combined to a single circuit using the programmable interconnect. The configurable interconnect is realized by regularly structured routing resources and configurable switch boxes, which combine different routing segments to entire paths that connect CLBs and/or coarse-grained blocks, such as memory blocks. Arbitrary hardware functions can be implemented on an FPGA using the configurable logic blocks and interconnect. The maximum clocking frequency of a circuit depends on the latencies of the input/output signals and on the circuit's critical path, which is defined as the longest path between two synchronous storage elements, such as flip-flops.

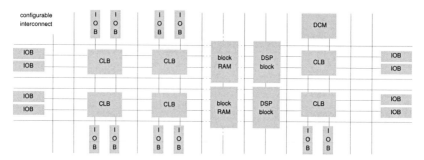

Figure 2.1: A schematic overview of an FPGA architecture (source: [152]).

Furthermore, FPGAs possess input/output blocks (IOBs), block memories (BRAMs[4]), digital signal processor (DSP) blocks and digital clock managers (DCMs). The IOBs are connected to the external pins of the FPGA, which are regularly distributed across the surface of the die. Dedicated logic inside the IOBs provides a wide set of input/output interface standards. The block memories are distributed across the area of the FPGA to guarantee a fast connection to a memory block at each location of the chip. Complex arithmetic operations such as a multiplication require a large amount of programmable logic resources. Hence, modern FPGAs contain digital signal processor blocks, which usually provide multiply-accumulate operations. Similar to BRAMs, DSP blocks are also distributed across the chip. Various clocking frequencies can be generated using the digital clock managers. Finally, some FPGAs provide further coarse-grained modules such as a hard-core processor block, which is integrated into the FPGA fabric. A schematic overview of an FPGA architecture can be seen in Figure 2.1.

Hardware circuits are programmed using a hardware description language (HDL). The source code is translated to the register-transfer level (RTL) by a synthesis tool. In the technology mapping phase, the RTL components are mapped to the physical components of a specific FPGA. The physical components can differ between FPGA families and packages, because not every FPGA has the same amount of DSP blocks or BRAMs. Furthermore, over the years the LUTs have increased their complexity and now support six instead of four inputs. The result

---

[4]block random-access memory (BRAM)

of the mapping phase is a netlist that is bound to a specific FPGA family. In the placement phase, the logic components of the netlists are placed to specific physical components of the FPGA. The placed components are then connected with each other in the routing phase. Certain heuristics are used for the placing and routing phases in order to fulfill the user-defined timing constraints for all circuits. Finally, a bitstream is generated that contains all the configuration bits of the logic components and the interconnect. The bitstream format of commercial FPGAs is confidential in order to to keep certain implementation details secret and to enforce designers to use the vendor tools.

Entire computing systems can be implemented on a single chip using **platform FPGAs**. Next to the FPGA, FPGA boards include further components such as DDR-SDRAM[5] modules and many input/output ports (Ethernet, USB[6], video, audio, etc.). Certain parts of the computing system, such as the processor, the memory management unit, system buses, Ethernet media access controllers, and many more, are implemented inside the configurable logic of the FPGA. Each of these system components can be adjusted to the needs and the desires of the system designer. This enables researchers to implement prototypes of novel computing architectures onto FPGAs. As the configurable logic and interconnect can be reconfigured arbitrary often, a single platform FPGA can be used to implement various prototypes of computing architectures over time. The configuration of an entire FPGA can take milliseconds to seconds depending on the size of the FPGA.

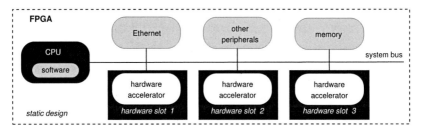

Figure 2.2: Reconfigurable systems-on-chip: digital components of a computer system are implemented in reconfigurable hardware

Computing system designers can additionally make use of the remaining unused reconfigurable logic resources to implement (application-specific) hardware accelerators. Application fragments that show a high degree of parallelism are a natural fit for a hardware implementation. However, as the resources of an FPGA

---

[5]double data rate synchronous dynamic random access memory (DDR-SDRAM)
[6]universal serial bus (USB)

are limited, not all hardware accelerators may fit on the FPGA fabric at the same time. For this purpose, most Xilinx FPGAs support dynamic partial reconfiguration where only certain regions of the FPGA fabric are reconfigured while the other parts of the system continue execution. To allow for partial reconfiguration the designer must divide the FPGA design into a static part and into dynamically reconfigurable regions. Then the designer builds a full bitstream for the static part of the system and partial bitstreams for the dynamically reconfigurable regions (hardware slots). The communication between the static and dynamic parts of the system design has to be well-defined. The partial reconfiguration can be accomplished externally over the JTAG[7] interface and internally using the ICAP[8] controller. This empowers reconfigurable systems-on-chip (rSoC) to have more hardware accelerators available than systems-on-chip that do not support run-time reconfiguration. An exemplary overview of a reconfigurable systems-on-chip is depicted in Figure 2.2.

2.2 Hybrid Multi-cores

In the last decade many prototypes were released that combined a processor with (reconfigurable) hardware accelerators. In this section, we present the related projects Molen, BORPH, DREAMS, Hthreads and RAMPSoC. These projects have in common that they have developed an operating system for a reconfigurable multi-core architecture, which consists of processors and hardware cores. The hardware accelerators are either implemented as processes or threads. All presented projects are compared to the ReconOS programming model and execution environment that is used in this thesis.

Molen

The Molen project covers several fields such as computing system architectures, compilers and run-time management techniques. The Molen polymorphic processor tightly couples a general-purpose processor with several reconfigurable custom-computing units [147]. By extending the instruction set of the processor by only eight additional instructions the processor is able to reconfigure its co-processors at run-time. An arbiter decides if an instruction is executed on the processor itself or on a reconfigurable co-processor. The first Molen prototype did not support multithreading. Later, Uhrig et al. [145] combined a simultaneous multithreading (SMT) processor with a reconfigurable co-processor. Although multithreading was supported on the SMT processor, only one thread could make

---

[7]joint test action group (JTAG)
[8]internal configuration access port (ICAP)

use of the co-processor at each time. However, threads could not be preempted during execution.

Furthermore, the Molen project also introduced compiler extensions in order to support a parallel execution on the reconfigurable co-processors for OpenMP-based applications [130]. Therefore, Sima et al. proposed an integer linear programming mapping algorithm that maps parallel applications to the processors in order to reduce the execution time of the application. The mapping is constrained by the fixed reconfigurable area of the co-processor. In [129], Sima et al., proposed a dynamic mapping algorithm that determines if a specific function should be executed in software or in hardware based on the current parameters of the function. Mushtaq et al. [106] presented a run-time profiler that collects statistics of code segments with a low overhead of 1.5% (in execution time). Finally, Mariani et al. [98] took advantage of profiling information gained by software profiling and hardware synthesis. In a multi-objective design space exploration the trade-offs between the area utilization of the reconfigurable fabric, the CPU load, and the application's performance are computed for different mappings. This information is used by a run-time manager that has to map multiple applications to the Molen architecture.

In strong contrast to the Molen project, ReconOS promotes passive co-processors to active hardware threads that can independently communicate and synchronize with other threads and can access shared system resources. Furthermore, we did not modify the compiler or the instruction set of our employed processors. Finally, our proposed run-time techniques do not require any results obtained by software simulations or hardware synthesis.

BORPH

One of the first approaches to provide a unified execution environment for software and hardware processes using a well-known interface is the Berkeley operating system for reprogrammable hardware (BORPH) [138, 139]. The conceptual overview of BORPH is depicted in Figure 2.3, where software and hardware processes can interact with the same operating system kernel. For the first prototyping platform the standard Linux kernel 2.4.30 was extended to support hardware processes. So-called BORPH object files store information about the configuration of hardware processes along with the corresponding bitstreams. The user can execute a hardware process similar to an executable file. The status of a hardware process can be observed using the ps command of Linux and the process can be terminated using the kill command. This simplifies the interaction with hardware processes for a software engineer.

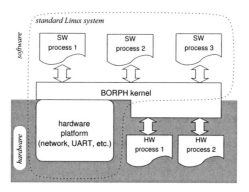

Figure 2.3: Conceptual overview of a BORPH system: Hardware and software processes have the same input/output interface. The dotted line encircles the standard Linux system (source: [138]).

A special feature of BORPH is the run-time file system support for hardware processes [87]. In the first implementations of BORPH, the communication between hardware and software processes is realized using the `ioreg` interface of Linux. The processes can read and write to virtual `ioreg` files, which might be implemented as FIFOs[9], registers or off-chip/on-chip memory. The first BORPH system was implemented on the BEE2 hardware platform, which consists of five Xilinx Virtex-II FPGAs. One central FPGA controlled the other four user FPGAs. The BORPH operating system was executed on an embedded hard-core processor on the central FPGA and the hardware processes were programmed to the user FPGAs. The operating system handled the interrupts of the hardware processes and used message passing for communication.

Recently, Changqing et al. [36] introduced shared memory support to BORPH where a workstation CPU[10] shares its main memory with a platform FPGA. The FPGA is connected to the workstation over the PCIe[11] interface. The FPGA contains a memory management unit, which translates virtual to physical addresses.

Similar to BORPH, our programming model uses a well-defined communication interface and provides shared memory. Unlike BORPH, we follow a multithreading approach and provide the well-known POSIX API for hardware threads. Furthermore, our execution environment ReconOS runs in the user-space. Hence, in contrast to BORPH, ReconOS does not require kernel modifications.

---

[9]first-in first-out (FIFO)
[10]central processing unit (CPU)
[11]peripheral component interconnect express (PCIe)

DREAMS

The distributed real-time extensible application management system (DREAMS) library provides skeletons for a customized distributed operating system [116]. In DREAMS, the optimal hardware/software mapping can be determined at run-time. Hence, the operating system can dynamically optimize the mapping of its operating system functionalities according to the application's requirements. More specifically, the operating system migrates its services in order to offer the currently most important system resources, i.e. a hardware accelerator, to the applications. A heuristic based on ant colony optimization was used to dynamically re-map the operating system services to cores in order to minimize the communication overhead.

Götz et al. [56–58] developed real-time scheduling techniques for periodic and aperiodic operating system services that can be either mapped to software or to hardware. A common task structure description for the software and hardware implementation was presented together with a complete tool flow that generates a software executable and a bitfile. The generated files were created for the DREAMS operating system. Task migration was supported using manually defined migration points.

In contrast to DREAMS, ReconOS is a monolithic operating system that does not support real-time constraints. We use a common programming model based on multithreading, which provides a common POSIX interface to the programmer. Furthermore, our mapping algorithms are triggered by sensors and base their mapping decisions on internal models.

Hthreads

Hthreads seeks to make complex reconfigurable devices more accessible to regular programmers [5, 113]. In Hthreads, software and hardware threads can access a comprehensive set of operating system primitives that are implemented as hardware modules. Hthreads supports individual hardware cores for thread management, scheduling, mutex-based synchronization, condition variables, and, interrupt scheduling. Accessing any of these operating system functionalities provides only a tiny overhead because, on the one hand, the hardware modules run in parallel, and, on the other hand, the hardware modules were designed to provide the service with a minimal latency. Each hardware thread has a well-defined interface to all system-level interactions and system calls. An example Hthreads architecture with one processor and two hardware threads is depicted in Figure 2.4.

Furthermore, in [22] Agron et al. showed that the Hthreads architecture can also be used for heterogeneous multi-processor systems where the processors differ in their instruction sets. As the operating system is implemented as a set of low overhead hardware cores, there is no central unit, which processes all operating system calls. Correspondingly, experimental results show that their decentralized operating system scales better than a centralized one. Recently, Cartwright et al. [33] proposed a tool flow that creates heterogeneous multi-processor systems-on-chip inside the cloud. Different architectures can be configured using a simple web interface. The threads are programmed using the standard `pthread` programming model and can be compiled for different architectures.

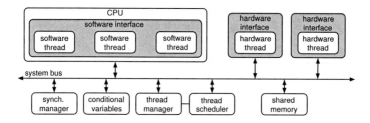

Figure 2.4: Hthreads system block diagram (source: [33]).

Both approaches, Hthreads and ReconOS, use a common multithreading programming model based on pthreads to allow programmers to code applications for hybrid architectures that consist of heterogeneous processors and (reconfigurable) hardware threads. Unlike Hthreads, ReconOS extends a monolithic operating system kernel, which runs on a central processor. On the one hand, ReconOS provides higher latencies for system calls than Hthreads and does not scale as well as Hthreads. On the other hand, ReconOS shows a greater flexibility, because ReconOS can extend any (open source) operating system kernel, which supports software multithreading. This enables ReconOS systems to take advantage of the different strengths of various operating systems, such as the rich feature set of Linux or the driver support for eCos. However, in Hthreads there is no concept for preempting hardware threads at any time, whereas in ReconOS cooperative multithreading supports preemption at predefined yielding points. Finally, Hthreads does not contain sensors to capture the on-chip temperature distribution and has not been investigated for dynamic thermal management techniques.

RAMPSoC

RAMPSoC stands for a runtime adaptive multi-processor system-on-chip where the hardware architecture can be modified at run-time [54]. The architecture contains several reconfigurable instruction set processors (RISPs) on a single chip. A RISP combines a standard processor with one or multiple reconfigurable hardware accelerators (co-processors). An adaptive switch-based network-on-chip (NoC) connects multiple RISPs with each other. In RAMPSoC, the hardware accelerator/s, the entire soft-core processor and also the communication infrastructure can be adapted using partial reconfiguration in order to increase the performance or to save power. For instance, the hardware accelerator is reconfigured if the application needs to accelerate a different function. When the data bit-width that has to be processed changes over time, the entire processor might need to be exchanged with another kind of processor. This can happen when a different data accuracy is required. For this case also the networking infrastructure needs to be updated to increase the data bandwidth. To save power, network switches that are currently not needed by the system can be removed or at least turned off. An example RAMPSoC architecture with three RISPs and one FSM-controlled hardware function is depicted in Figure 2.5.

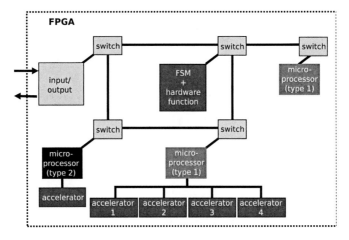

Figure 2.5: Example RAMPSoC architecture (source [55]).

The first RAMPSoC approach did not mention active hardware processes. Instead, hardware accelerators were integrated as passive co-processors. However, in later publications [52, 55], Göhringer et al. also introduced hardware accelerators that are controlled using a final state machine (FSM). In [55], the authors propose an operating system called CAPOS for runtime scheduling, task mapping and resource management. The configuration access port-operating system (CAPOS) manages the run-time reconfiguration of the RAMPSoC architecture. This is challenging because current Xilinx FPGAs only allow the usage of one ICAP reconfiguration interface at the same time. The operating system runs on one of the processors that is directly connected to an ICAP interface and the external memory that stores the partial bitstreams of every task. CAPOS bases on the Xilkernel [163] and schedules an application that is given as a task graph. Whenever CAPOS cannot assign a ready task to a processor it analyzes if a new processor should be configured to the system. The reconfiguration can only be performed if there is enough free space left on the FPGA. Furthermore, [52] presented an adaptive multi-client network-on-chip memory that provides a dynamic mapping of address space for the different clients within the network. Finally, [53] presented two different system monitors that were used for dynamic frequency scaling. One monitor captured the processor load and the second the communication load of the network.

The RAMPSoC approach offers a more advanced system architecture and more types of run-time reconfiguration compared to ReconOS. In contrast to ReconOS, which uses a system bus, a network-on-chip is used in RAMPSoC. However, ReconOS provides a well-defined multithreading interface (POSIX) to reconfigurable hardware threads. Furthermore, thread migration has not been studied as a run-time technique for performance and thermal management on RAMPSoC. Finally, RAMPSoC does not include temperature sensors in their architecture and does not support virtual memory.

2.3 Task Migration across the Hardware/Software Boundary

To apply thread migration techniques to reconfigurable devices such as FPGAs, different context save and restore approaches were researched. Simmler et al. [131] first proposed a technique for transparent context saving and restoring by bitstream read-back and manipulation. By refining these concepts, Kalte and Porrmann [76] implemented an architecture for relocatable hardware tasks, which allows the extraction of state values from an FPGA's storage elements and their injection into partial bitstreams for reconfiguration in a different location on the device. Their approach is transparent to the hardware module's designer, but involves considerable time overheads. The complete bitstream of the reconfigurable area must be read back and processed, although usually less than 1% of the configuration data contribute to the actual module's state. In addition, the bitstream format is usually not open and must be reverse engineered for every FPGA device family. As an alternative approach, Jovanovic et al. [75] as well as Koch et al. [83] proposed linking registers together in a serial scan-chain that can be used to read or write a hardware module's context in a transparent manner. These scan chains can be automatically inserted on the source code or netlist levels. They do, however, incur a significant time and area overhead.

The preemption and migration of tasks across the hardware software boundary was investigated, for example, by Götz et al. [57]. They proposed a framework that generates templates for hardware and software tasks based on a task structure description. These templates are then filled out by the task designer, including a specification of viable migration points. Mignolet et al. [102] implemented a comparable thread migration technique by integrating checkpoints at which task interruption and context transfer is possible. They generate both hardware and software task representations from the same description and map inter-task communication to a message-passing mechanism. Koch et al. [82] modeled the migration of FSM-based hardware tasks to software and vice versa on predefined morph points, where equivalent context representations exist in hardware and software together with a mapping morph function. They automatically insert the state access logic necessary to access a hardware task's state by a custom tool chain. A formal investigation of hardware/software state equivalence, motivated by the migration of applications between software and FPGAs, was presented by Blumer et al. [29]. The authors developed a taxonomy of migration realms consisting of processors modeled as finite state machines and describe the migration of processes within a migration realm.

Pellizzoni and Caccamo [114] proposed a real-time computing architecture that supports quality-of-service (QoS) adaptation on a hybrid CPU/FPGA system featuring seamless migration of (periodic) tasks between hardware and software. A hardware task runs for the entirety of its period. Thus, at most one task at a time is allocated to a hardware region. Since there is no support for task preemption, migration can only occur at the start of a period. Additionally, a task exists in multiple configurations (variants) for different QoS constraints, typically a software version and several differently optimized hardware implementations. Tasks are migrated in response to newly incoming tasks, changing QoS constraints, or differing workloads.

While in principle our approach to hardware/software migration is similar to [57, 82, 102], the main difference is that the thread migration transparently integrates itself into the multithreaded programming model and operating system. The automated specification of suitable migration points from a common hardware/software specification is not part of this thesis. For streaming applications, which are natural workloads for hybrid multi-cores, the manual specification of migration points is not too difficult.

2.4 Performance Management

This section provides related work on performance monitoring, performance models and performance management techniques that rely on thread migration.

Performance Monitoring

Twenty years ago, performance monitors were already used to gain run-time information about programs. For instance, Harden et al. [65] proposed a hybrid performance monitoring infrastructure for multi-computers where a performance monitor is placed at each compute node. Whenever objects are created, messages were sent or received, operators are started or stopped, this information is logged together with a time stamp in the memory. All monitors were connected to each other in a tree-like structure and a central computer, which was attached to the multi-computer, collected and displayed all data. Kim and Kim [81] discussed the need of performance monitors for single-chip multiprocessor digital signal processors. Three performance monitors were proposed: one for custom monitoring, one for cache monitoring, and one for contention monitoring. The cache monitor collected the number of cache misses and hits, the average service time of a cache miss and further cache-related information. The contention monitor captured the total number of crossbar switch contentions and the custom monitoring could be used to log the executions of checkpointing instructions.

More recently, Hatzimihail et al. [66] detected so-called performance faults that can occur after an incorrect branch prediction or a wrong data value prediction using a performance monitor. Here, the term 'performance faults' does only refer to a miss-speculation of the corresponding hardware module, which is in charge of prefetching instructions or data. Faulty speculations are corrected by the module itself. Hence, faulty speculations do not propagate to actual faults in functionality but they degrade the performance of the processors. The authors used performance counters to monitor faulty branch predictions and modified the branch prediction strategy at run-time in order to improve the performance of a RISC[12] processor. Targeting super computers, Salapura et al. [120] developed a complex performance monitoring unit, which was able to watch thousands of concurrent events simultaneously. In the proposed monitoring architecture, the least significant bits of the counter were stored in registers and the most significant bits in a dense SRAM[13] array. Furthermore, the events could send interrupt signals whenever they reached predefined thresholds.

---

[12]reduced instruction set computing (RISC)
[13]static random access memory (SRAM)

The Heartbeats framework, developed by Hoffmann et al. [67], is an infrastructure to monitor the performance of applications. There, every application must send a signal, i.e. a heart beat, periodically to the Heartbeats framework. The heart rate of an application provides information about its current performance. The performance goals are given by the application-specific heart rates, i.e., heat beats per second. For a streaming application, a heart beat can be send whenever a data packet has been processed. Our performance monitor is similar to the Heartbeats framework, because we monitor the performance of streaming applications in general and the execution time of specific threads in particular. However, in contrast to Heartbeats our performance monitoring is integrated into the application.

Performance Models

Estimating the performance becomes challenging due to the increased complexity and heterogeneity of today's embedded system. Accurate performance estimation can be used at design-time to create the 'golden' architecture [111], which fits best to the target application. However, due to run-time dynamics, performance estimations can also be used to drive the mapping of tasks to cores at run-time. Many researchers presented novel approaches and tools for performance estimation and analysis. For instance, Oyamada et al. [111] presented a performance estimator based on neural networks. Compared to a cycle-accurate virtual platform simulation, the presented estimator provides a speedup of 35 while suffering an estimation error of 17% when it is applied to an MPEG[14]-4 encoder.

Cheung et al. [39] proposed structural performance models that estimate the software performance on multiprocessor systems-on-chip (MPSoCs). The authors modified the machine description of the GCC[15] compiler. Thereby the structural models could be automatically generated by the GCC compiler. The performance estimation error is below one percent compared to a cycle-accurate instruction set simulation, while providing a speedup of several orders of magnitudes at the same time. Instead of simulations, formal models can also be applied for performance analysis. For instance, Huang et al. [70] presented a formal performance analysis model for system validation. The proposed model targets streaming applications, which are given as data flow process networks where autonomous actors communicate with each other using FIFO queues.

Our performance model does not include complex performance estimation methods or formal models for a performance analysis. Instead, the execution times of selected threads of a streaming application have been benchmarked for all core

---

[14]moving picture experts group (MPEG)
[15]GNU compiler collection (GCC)

types in order to determine the speedup values for all (core, thread) combinations. This information is already given in the performance model. In this thesis, we assume that the system has already learned this information.

Thread Migration for Performance Management

In recent years, thread-level adaptability of embedded devices in both micropro-cessor and reconfigurable hardware contexts have been investigated from different angles. For example, Kumar et al. [85] argued that for many applications, core diversity is more important than uniformity to better adapt to the changing demands of applications. The authors designed and simulated a heterogeneous multiprocessor system where the processors show different power/performance characteristics. During the application's execution the system software selects the most appropriate core that saves the most energy while still achieving a specified performance. For 14 SPEC[16] benchmarks, they achieved an average energy saving of 39% with a performance loss of 3%.

Curtis-Maury et al. [42] recognized that threads targeted at complex multi-core systems interact in a complex and data-dependent manner and proposed to adapt the thread / core distribution based on run-time thread execution profiles. They developed a multithreading library for power / performance adaptation and evaluated their approach on OpenMP-based applications running on a symmetric multi-core system without heterogeneous hardware cores.

Huang and Xu [71] studied process variation of MPSoCs that can result in variations in frequencies and leakage powers among the processors on the same chip. Based on the actual process variation of the chip, the proposed methodology selects the appropriate task schedule out of a set of variation-aware schedules that were computed at design-time in order to improve the probability to meet all timing constraints.

In the field of dynamic task mapping in FPGA-based systems, Stitt et al. [141] proposed to dynamically partition applications between software processors and hardware co-processors, where co-processors are generated at run-time using binary decompilation and synthesis. Huang and Vahid [69] showed that the dynamic co-processor management can be reduced to the metrical task system problem and presented a fading cumulative benefit heuristic. The heuristic stores the theoretically achievable cumulative benefit for each application that could have been achieved by using a co-processor. The fading cumulative benefit values are fading in order to focus on temporal locality. Sigdel et al. [128] introduced a two-level design space exploration (DSE) to solve the dynamic task

---

[16]standard performance evaluation corporation (SPEC)

mapping problem. At design time, the first level DSE explores the system under static conditions to find the optimal static task partitioning. At run-time, the second level DSE seeks to optimize the task mapping to adapt to changes in the application, architecture, or the environment.

Similar to [69, 128], we map tasks to processors and reconfigurable logic. In contrast to Vahid et al., who generate hardware co-processors from parallelizable code, we map entire threads (including operating system calls) into hardware. While most approaches try to maximize the performance, we aim to achieve a desired performance while minimizing the number of active cores.

Targeting more heterogeneous systems that also comprise DSPs, ASICs[17], etc, Smit et al. proposed two heuristics, which minimize the energy consumption: the adapted minWeight heuristic [137] and the more advanced hierarchical iterative heuristic [136]. Moreover, several heuristics for NoC-based systems were published, i.e. [34, 107]. Nollet et al. [107] improved the task assignment success rate by dynamic instantiation of soft-core processors that allow tasks, which do not have a hardware implementation, to be mapped to the FPGA fabric. In contrast to our work, these approaches either focus on energy consumption or on the properties of NoC-based architectures.

2.5 Thermal Management

Continuously shrinking microelectronic device structures lead to novel challenges in complexity, productivity, and reliability. The reliability challenge is posed by increasing variations in device behavior, single-event upsets, and device degradation, as described by Borkar [31]. Several authors analyzed the different physical sources for the variability and degradation of semiconductor components, e.g., Unsal et al. [146]. Rivers and Kudva [117] discussed reliability challenges at the architecture level and provided a comprehensive classification of physical flaws, errors, and failures. Furthermore, the authors made the case for hybrid systems and claim that software-level techniques to provide reliability will be of increasing importance in the future.

For SRAM and eDRAM[18] cache memories, Meterelliyoz et al. [100] showed that temperature variations lead to instabilities. As FPGAs hold their configuration bits in SRAM cells, these results also apply to FPGAs. Seyab and Hamdioui [126] analyzed the temperature impact on various sub-processes that contribute to negative bias temperature instability (NBTI) degradation. According to Wang et al. [150], NBTI induced degradation is relatively insensitive to supply voltage, but

---

[17]application specific integrated circuit (ASIC)
[18]embedded dynamic random access memory (eDRAM)

strongly dependent on temperature during dynamic operation. Many approaches were presented to combat the effects of variation and degradation on technology, circuit, and micro-architectural levels, including self-timed asynchronous design styles, variation-aware design methods, error checking codes, and redundancy techniques.

Temperature Monitoring

Proactive thermal management techniques seek to prevent thermally induced degradation and errors. However, since today's FPGAs only contain a single thermal diode, many research projects implemented additional temperature sensors into the reconfigurable logic in order to gain more insight to the current temperature profile of the chip.

Ring oscillators are widely used as temperature sensors in FPGA-based systems. For instance, Lopez-Buedo et al. [89] showed that the frequency of a ring oscillator is inversely proportional to the temperature. Their sensor consists of a ring oscillator with seven inverters, a timebase counter and a capture counter. For a temperature measurement, they initially enabled the ring oscillator for 1024 clock cycles before they performed the measurements for 2048 clock cycles. Velusamy et al. [148] additionally validated their ring oscillator-based temperature sensors with HotSpot [73], a popular thermal modeling tool for VLSI[19] systems. In [89, 148] the sensors were calibrated using a temperature-controlled oven.

Zick and Hayes [169] developed an efficient implementation of a ring oscillator-based sensor that can be used for delay, static power, dynamic power, and temperature measurements. Sayed and Jones [123] characterized non-ideal impacts of reconfigurable hardware workload on ring-oscillators on Xilinx Virtex-5 FPGAs and showed that the workload has to be considered to be able to use ring oscillators as temperature sensors.

Besides ring oscillators, infrared cameras gained importance for temperature measurements. Nowroz et al. [108] devised techniques that fully characterize the thermal status of an Athlon dual II processor using a limited number of measurements from thermal sensors and comparing the temperature characterization with an infrared camera. Nowroz and Reda [109] proposed so-called soft sensors for thermal and power characterization of FPGAs using infrared imaging techniques. A soft sensor measurement is equal to a weighted linear combination of the measurements of real sensors. The sensor weights can be calibrated using an infrared camera. The technique reduces the required real sensor resolution

---

[19]very-large-scale integration (VLSI)

on the device. Generally, the use of thermal models in addition to temperature sensors can reduce the number of required sensors [127].

In this thesis, we use ring oscillators combined with counters as temperature sensors, similar to [169]. In contrast to less recent related work [89, 109, 148, 169], we do not calibrate the sensors using a temperature-controlled oven or an infrared camera. Instead, our system self-calibrates its sensors using local heat-generating cores and an internal thermal diode, which can be found in modern FPGA devices, e.g., Xilinx Virtex-5 and Virtex-6 FPGAs.

Thermal Models

HotSpot [72, 73] is a widely-used tool for thermal simulation of VLSI designs. HotSpot provides a modeling methodology that makes use of the duality between thermal and electrical phenomena and defines a multi-layered RC[20]-network to compute the heat flow on the chip. The model comprises multiple layers such as a heat sink, heat spreader, thermal interface material, silicon bulk, interconnect layer, etc. A simplified floorplan of a typical VLSI package is shown in Figure 2.6. The integrated circuit (IC) die is embedded inside an IC package. The IC package is attached to a printed circuit board (PCB) using several package pins. A heat spreader transfers the heat of the die to a heat sink, which is placed on top of the heat spreader.

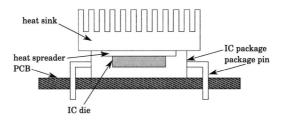

Figure 2.6: Overview of a typical VLSI package (source: [135]).

Next to HotSpot, further thermal models have been proposed to simulate the heat flow for multi-core processors [47, 48, 110]. Zhang et al. [167, 168] addressed the problem of estimating the chip-level thermal profile at run-time using only a few on-chip sensor observations and random chip power density characteristics due to unpredictable workloads and fabrication variability. The authors improve their estimation accuracy by exploiting the correlation between temperature and power density.

---

[20] resistor-capacitor (RC)

Hanumaiah and Vrudhula [63] used accurate power and thermal models to provide a so-called reliability-aware thermal management for hard real-time applications on multi-core processors. They model the task of finding optimal core speeds for minimizing the peak temperature under given task deadlines as a quasi-convex optimization problem and report on a reduction in peak temperature of 8°C for the SPEC benchmarks. Wang et al. [149] proposed the FRETEP method to accurately estimate and predict the full-chip temperature at run-time relying on a limited number of built-in thermal sensors as well as inaccurate thermal models and power estimations. The authors use RC-networks to model thermal behavior, similar to HotSpot, and devise an error compensation method based on thermal sensors. This thermal model is reported to accurately estimate and predict the temperature distribution at run-time with a low overhead; in particular it is found to be superior to a Kalman filter based approach on the standard SPEC benchmarks.

Our thermal model uses an RC-network approach and is thus in line with HotSpot [72, 73] and other related work [63, 149]. However, we only model two layers trading modeling accuracy for performance. Also, in contrast to [72, 73] we do not assume knowledge of material properties provided by the manufacturer to define model parameters but learn those parameters on-line.

Thread Migration for Thermal Management

Over the last years, several dynamic thermal management techniques were discussed, which aim at avoiding temperature hot spots (locations on the chip where the temperature exceeds a certain temperature threshold) and at balancing the on-chip temperature [32, 41, 49, 84, 105, 148]. Since thermal management is closely related to power management, we find dynamic frequency scaling (DFS) and dynamic voltage scaling (DVS) among the main techniques for dynamic thermal management. For example, the Pentium 4 processor provides an on-die thermal diode [14] and software-controlled clock modulation [15], which allows the operating system to throttle the processor clock in discrete steps if the temperature exceeds a certain threshold. This section focuses on thermal management, which is achieved by migrating threads between different cores of the system.

Activity migration is a means to decrease temperature by spreading electrical activity over a larger area. At the level of multi-cores, activity migration means migrating threads from core to core. Michaud et al. [101] studied thread migration in temperature-constrained multi-cores based on a rather simple on-off (also denoted as global clock gating) thermal management strategy for cores.

Kursun and Cher [86] discussed two schemes for dynamic variation-aware thermal management termed core hopping and thermal-aware task scheduling. They generated a variation map for individual cores from on-chip thermal sensors. To that end, they ran special benchmarks for stressing the cores for characterization, and calculated the deviation of the core temperature from the chip average, respecting the start temperature and capturing the core's activity counter. The cores were then ranked by their criticality, where higher core temperatures relate to higher criticality. The three parameters temperature deviation, activity count, and core criticality form the core's variation coefficient stored in the thermal variation map. Thermal-aware task scheduling uses the thermal variation map to simply assign tasks to suitable cores, whereas core hopping means to move computation from hotter to colder cores. Experiments performed on an IBM 1.5 GHz test chip with a 4-core simultaneous multithreading processor running Linux and hopping periods in the 10-100 ms range resulted in peak temperature reductions of some degrees for parts of SPEC2006 without compromising performance.

Yeo et al. [165] proposed predictive thermal management to estimate future core temperatures based on two models, an application-based model for short-term thermal behavior and a core-based thermal model for long-term thermal behavior. The predicted temperature is a combination of both temperature models, where the authors suggested a 70% weight for the short-term component. The proposed approach used certain temperature thresholds to trigger thread migrations. Here, threads were migrated from cores that violated the temperature thresholds to cores that were predicted to have the lowest core temperature in future. Yeo et al. also studied temperature profiles and observed that the temperature slope depends on the difference between current temperature and the steady state temperature of the application. The steady state temperature would be achieved when the application runs forever on the core. An extended Linux scheduler decreased the average core temperature by some 10%, and the peak temperature by 5°C over a standard Linux scheduler with negligible performance overhead on the SPEC2006 benchmark.

In a next step, Yeo and Kim [164] measured thermal profiles and approximated steady state temperatures for different applications and every core in the system. Then, they formed thermal behavior patterns from the temperature profiles, which are divided into steep, gentle, and flat regions, and apply k-means clustering. When running a new application on a core, its thermal behavior pattern is predicted by classifying the best matching cluster according to the current temperature and slope. Subsequently, the scheduler runs the application on the core that needs the longest time period to reach the desired temperature threshold. Yeo and Kim implemented their scheme on Intel quad cores and reduced the peak temperature over the standard Linux scheduler by up to 8°C at 12% performance overhead on one quad core, and by up to 5°C at some 8% performance overhead on two quad cores.

In the domain of MPSoCs, Mulas et al. [105] presented a dynamic temperature-aware scheduling policy called MiGra, which bounds temperature gradients using task migration. At a lower level, each processor is controlled by dynamic voltage/frequency scheduling depending on the current workload. At the higher level, an advanced scheduling algorithm migrates tasks from processors exceeding an upper temperature threshold to processors whose temperature is below a lower threshold. Both thresholds are dynamic and depend on the average chip temperature. MiGra selects a pair of processors for task set migration according to migration costs that include estimates of performance, energy, and temperature increase. MiGra balanced the temperatures of the processors within a range of 3°C around the average temperature with a performance overhead of 2% for stream computer platforms using three cores.

Eisenhardt et al. [49] presented an activity migration technique for coarse-grained reconfigurable architectures that aims at preventing hot spots by continuous row-wise activity migration. Exactly one row of elements in the array is active for a fixed time interval, before the activity is migrated to the next row, rolling over at the final row. Although the compute elements in this architecture are considerably smaller than fully fledged processor cores, this thermal management technique incurs a relatively high overhead in required compute elements compared to an architecture that does not perform activity migration and, therefore, does not contain redundant rows of compute elements. For instance, when an architecture contains r rows, the presented approach imposes an overhead of $r - 1$ rows of additional compute elements. Moreover, the technique can be classified as static since the migration schedule is known a priori. In the area of fine-granular reconfigurable devices, Gupte and Jones [60] proposed the idea of thermally-driven task swapping between hot and cold hardware modules using partial reconfiguration. The technique bases on thermal sensors, which are placed inside the hardware modules.

Ge et al. [51] applied a distributed task migration technique on many-core systems. A balanced thermal profile could be achieved by proactive task migration among neighboring cores. A low-cost agent resides in each core and observes the local workload and temperature, and communicates with its nearest neighbor for task migration. Compared to a predictive dynamic thermal management (previously proposed by Yeo et al. [165]), the proposed thread migration policy is reported to reduce thermal gradients by 66.23% and hot-spots by 66.79%, and to improve the performance by 40.21% while maintaining a 33.84% lower migration overhead.

Hanumaiah et al. [64] presented thermal management techniques for multi-core processors, including both DVFS and task migration. Their simulation results show that their task-to-core allocation technique provides a 20.2% improvement in performance over a power-based thread migration approach. HotSpot has been used for thermal modeling. Wu et al. [153] proposed a scalable distributed thermal management technique for a mixture of workloads consisting of deadline-driven and general-purpose tasks. In an initial step, the thermal correlation between the cores is quantified. This information is then used at run-time when a temperature of a core reaches a certain threshold. The distributed controllers of thermally correlated cores communicate to determine the best core for a slow down. The authors showed that their so-called DistriTherm technique could reduce the deadline miss rate by 47.16% on average while achieving the comparable reduction of the peak temperature like fully localized thermal management techniques.

Ebi et al. [46] devised a power budget economy strategy for thermal management in many-core architectures. They could reduce the peak temperatures by around 4% compared to a fully distributed approach by applying an economic learning approach. The results were verified using an infrared camera. Recently, Cher and Kursun [38] introduced a technique that leverages the on-chip sensor infrastructure as well as the capabilities of power/thermal management to reduce heating and minimize local hot-spots. The proposed technique generates a variation map for individual cores and uses this map for accurate run-time estimation of the temperature distribution. For high-performance processors clocked at 1.5 GHz running Linux the authors have shown a reduction of the peak temperature by 5°C without any noticeable performance loss. Other related work targets three-dimensional multi-core architecture [61, 77, 97] using thermally-aware on-line task allocation and/or adaptation.

This thesis is totally in line with related work on thread migration, e.g., [38, 46, 51, 86, 101], with respect to the optimization goals (avoiding local hot spot, balancing the chip temperature). In contrast to related work, we deal with hybrid multi-cores, which leads to novel intra-modal (hardware-hardware) and even trans-modal (hardware-software and vice versa) migration opportunities.

2.6 Chapter Conclusion

First, this chapter established a background on field-programmable gate arrays. Due to fine-granular configurable logic elements and configurable interconnect, FPGAs are ideal platforms to implement novel computer architectures on a single chip. By taking advantage of the dynamic partial reconfigurability, configures hardware modules can be exchanged at run-time with hardware modules that are stored in external memory. Hence, hybrid multi-cores, which consist of multiple homogeneous or heterogeneous processors and reconfigurable hardware accelerators, can be implemented on a single chip.

Second, this chapter presented related research projects, which combine a hybrid multi-core architecture with an operating system. Over the years hardware accelerators were promoted from passive co-processors to active hardware threads. Furthermore, it can be seen that the trend goes towards NoC-based heterogeneous multi-core systems and towards distributed operating systems. This thesis takes up the trend of using active hardware threads instead of hardware co-processors, but does not follow the other two trends. However, the basic concepts and algorithms, which will be presented in this thesis, do not rely on the fact that our operating system is monolithic or that the cores are connected using a system bus. Hence, the results of this thesis can also be applied to related reconfigurable system-on-chip architectures with minor modifications.

Third, related work on performance management and thermal management was presented. Over the last years both run-time management techniques were studied intensively on heterogeneous multi-processor systems. Thereby, an increasing number of techniques employed task migration as tool of choice for run-time adaptations. Although many researchers targeted heterogeneous multi-core architectures, research on (self-adaptive) hybrid multi-cores, which additionally integrate reconfigurable hardware cores, is missing. Therefore, this thesis targets performance and thermal management techniques on hybrid multi-cores.

The next chapter will present our fundamental concept of using multithreading in hybrid multi-cores and will discuss the general ideas how thread migration can be applied by the system autonomously in order to adapt the application's performance and on-chip temperature distribution at run-time. Furthermore the next chapter will outline how internal sensors can be used to observe the system state and to detect time steps when thread migration is beneficial or even required. Finally, it will define the self-adaptation goals for performance and thermal management.

Alice had got so much into the way of expecting nothing but out-of-the-way things to happen, that it seemed quite dull and stupid for life to go on in the common way.

Lewis Carroll, *Alice's Adventures In Wonderland*

CHAPTER 3

Concepts and Key Ideas

This chapter presents the concepts and key ideas of this thesis. First, the chapter shows how the popular multithreading approach can be extended to reconfigurable hardware logic. The multithreading approach is used to build hybrid multi-cores. Furthermore, this chapter elaborates why a master-slave approach is used for operating such systems. Next, it explains how self-adaptation can be achieved using thread migration and how this technique can be used to manipulate the system's performance, i.e. the application throughput, and the on-chip temperature distribution. Moreover, this chapter identifies the target applications of these multi-cores that were covered in this thesis and defines certain requirements for the system architecture, such as the sensor layout for temperature measurements. Finally, this chapter formulates the goals for performance management and thermal management on self-adaptive hybrid multi-cores.

3.1 Multithreading in Hybrid Multi-cores

In the multithreading paradigm, applications are partitioned into several threads that access the same system resources such as memory, communication and synchronization primitives. The multithreading approach is widely-used in software-based systems. The greatest advantage of multithreading is that concurrent threads can be executed in parallel on different processors, which increases the performance in execution time of the applications. Multithreaded applications gained special importance, when the clocking frequency and the instruction level parallelism of processors could not be increased in the usual manner anymore and multi-processor systems were introduced to further increase the performance of a computing system.

The application developer can benefit from this partitioning as he can focus on well-defined partitions of the application. He can debug, validate and verify individual threads independent from the rest of the application. However, since the threads access shared resources, conflicts can occur when concurrent threads access the same resource at overlapping time intervals, e.g.., performing a write access to a shared memory region. These conflicts can be omitted when the threads use synchronization methods that guard the shared resources, e.g., by using semaphores or mutexes.

These synchronization primitives cannot prevent deadlocks between threads. For instance, two threads can lock the same two mutexes in a different ordering. This can result in a situation where both threads have locked one mutex and wait for the second mutex, which is locked by the other thread. Therefore, special verification techniques, applied at design-time, and resource reservation protocols, applied at run-time, exist to tackle this issue. In this thesis we assume that our applications are deadlock-free and all shared resources are well-protected using corresponding synchronization techniques. Threads can either communicate using shared memory or using dedicated communication primitives. For instance, POSIX provides an API that contains message queues for communication.

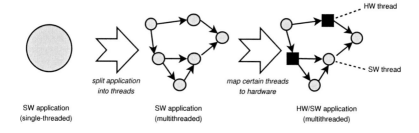

Figure 3.1: Multithreading concept: A single-threaded software application can often be divided into multiple threads that interact with each other. Some of these threads can be represented using hardware logic.

Figure 3.1 depicts the multithreading concept, which is used in this thesis. A single-threaded application is partitioned into multiple threads that communicate with each other. Multithreading is commonly used in software systems being executed on homogeneous multi-processor systems. In contrast to most related work, we can implement the threads not only in software but also in hardware. We use the ReconOS programming model and execution environment [95] that extends the multithreading concept to reconfigurable hardware. Here, passive hardware co-processors are upgraded to active hardware threads, which can access the same operating system functions as software threads. Therefore, ReconOS extends state-of-the-art multithreaded operating systems such as Linux (for high-performance systems) and eCos (for embedded systems) to support hardware threads.

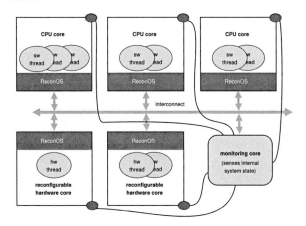

Figure 3.2: Hybrid multi-core architecture [11]

This thesis studies self-adaptive hybrid multi-cores that execute multithreaded applications. The hybrid multi-core consists of multiple processors that execute software threads and reconfigurable hardware cores that execute hardware threads. The multi-core has dedicated sensors for each core that capture the system state. For instance, thermal diodes measure the core temperatures and performance counters measure the execution time of specific threads. These core-specific sensor information are collected in a monitoring core and serve as an input for our internal system models.

It is the objective of a self-adaptive hybrid multi-core to autonomously manage one or several goals. If the system reaches an undesired system state or is about to reach such a system state, it is able to influence its state by adaptation. In this thesis we focus on thread-level adaptation, where we migrate threads between cores that possibly have diverse performance/thermal/power characteristics. Figure 3.2 shows an example hybrid multi-core that consists of three processors and two reconfigurable hardware cores.

In a hybrid multi-core we can differentiate between inter-modal and trans-modal thread migrations. For inter-modal thread migrations, the thread does not change its modality. This means that software threads can only be migrated between processors and hardware threads can only be migrated between hardware cores. Migrating a thread from a processor to a reconfigurable hardware core (or vice versa) requires to cross the hardware/software boundary and, thus, a change in modality of the thread. The trans-modal thread migration imposes a great challenge because the thread's context has to be translated between the different modalities. Figure 3.3 shows two examples for intra-modal thread migration (1-2) and one example for trans-modal thread migration (3).

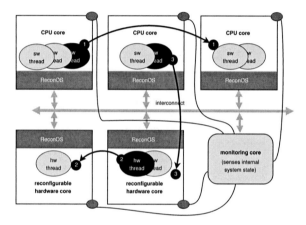

Figure 3.3: Supported types of migrations at a hybrid multi-core architecture: (1) intra-modal processor→processor migration, (2) intra-modal hardware core→hardware core migration, and (3) trans-modal processor→hardware core migration

3.2 Operating System for Hybrid Multi-Cores

According to [142] there are three ways to build operating systems for multi-processor systems. This section presents the three models and evaluates which model is best-suited for our self-adaptive hybrid multi-cores.

1. **Private operating system per processor:** Each processor executes its own operating system and the memory is divided such that each processor has its

own memory. The operating system code might be shared by all processors. This approach can also be seen as a cluster of individual computers.

2. **Master-slave multi-processors:** There is one master processor that runs the operating system and controls the other processors, which can only execute user processes and threads, i.e., applications threads. The master processor can execute both, the operating system, on the one side, and user processes and threads, on the other side. As there is only one operating system, there is no inconsistency between internal buffer caches of multiple operating systems anymore. However, this approach does only scale well for a limited number of processors, because the master processor has to perform the operating system calls for every processor.

3. **Symmetric multi-processors:** Here, the operating system is shared between all processors. The processors that performs an operating system call also processes it. The operating system is divided into several disjunct parts. Whenever a processor wants to make a system call of one operating system part, it has to lock it first, to avoid conflicts in accessing the shared resources, such as memory. This model assumes that each processor can execute each operating system part.

Using the first approach eliminates the opportunity to migrate threads between processors because the private operating systems do not cooperate among each other. Thus, we do not use this option. Furthermore, there can be inconsistencies between the buffer caches of the private operating systems. Finally, we do not intend to port the entire operating system to a hardware circuit for our reconfigurable hardware cores.

In this thesis, we consider multi-core systems with up to 15 cores. Furthermore, we consider applications, which usually spend most of their time in the processing of data and much less time in calling operating system functions (mostly for communication and synchronization). Hence, we believe that we can apply the master-slave approach for our hybrid multi-core systems. Finally, we only have to provide the operating system for one processor and not for all the other cores, such as the hardware cores, which eases the implementation and maintaining of the operating system. Thus, we use the master-slave approach for our hybrid multi-core.

We use the ReconOS approach, which already implements the master-slave approach for hybrid multi-cores. In contrast to previous work on ReconOS [90], we have ported the ReconOS system to modern FPGA technologies, added the opportunity to migrate threads between the hardware/software boundary and integrated monitoring units into the architecture. The principle of using the master-slave model remained the same.

For a hybrid multi-core, the implementation of such a shared operating system, which has been proposed for symmetric multi-processors, is complex because each hardware core needs to be able to execute the entire operating system. To port the operating system services entirely to a set of hardware circuits is highly complex and would result in a high overhead in memory (to store the OS hardware circuits) and in time (to reconfigure the operating system into the hardware slot). Due to the high reconfiguration times of reconfigurable hardware, a system call forwarding to a processor would be more efficient. Furthermore, updates in the operating system would require also changes in the hardware implementation of the operating system.

If we additionally consider heterogeneous processors with individual instruction sets, we would have to ensure that the operating system is available for each processor. Hence, a symmetric multi-processor (or symmetric multi-core) approach is inappropriate for hybrid multi-cores as long as the master-slave approach scales well. A possible solution would be to distribute the operating system not to all but a subset of cores (processors) that can handle all operating system parts. In this thesis, however, we apply the master-slave model.

3.3 Self-adaptation through Thread Migration

Most operating systems for single-processor systems provide preemptive multitasking. In preemptive multitasking a thread can be preempted at arbitrary points in execution and can be replaced by another thread. A scheduler decides which thread currently runs on the processor. A popular scheduling technique is the Round Robin time slicing technique, where the threads are stored in queues. If there is only one queue, the scheduler executes every thread in that queue for the duration of a time slice before it preempts it to replace it with the next thread in the queue. If the scheduler reaches the end of the queue, it starts at the beginning again. If a thread terminates, it gets removed from the queue.

If a thread A is replaced by another thread B, the context of the thread A has to be stored by the operating system. Furthermore, the context of thread B has to be loaded to ensure that the thread B resumes its execution correctly. The context of a thread is well-defined for software threads. The thread context contains the register values of the processor, the program counter and additionally information of the operating system, such as the process control block. The thread swapping times are usually negligible. Preemptive multitasking is popular because it allows the operating system to implement efficient scheduling algorithms.

The thread migration between two homogeneous processors C_1 and C_2 can be compared to a thread swap. The context of the thread being executed on core C_1 needs to be stored first. Then, in a second step, the thread context needs to be loaded onto core C_2 before it can resume. However, this preemptive technique can not be seamlessly applied for heterogeneous multi-processor systems. The processors might have different instruction sets and thus, a thread can not be migrated at arbitrary times.

Furthermore, the binary program code might be different for different processors. Thus, different processors might need to point to different memory regions for the program code and even alter the program counter. For instance, let us assume that we want to migrate between two heterogeneous processors C_3 and C_4 with different instructions sets. One processor C_3 is able to perform a complex instruction I, while the second processor C_4 needs to split this complex instruction into several instructions $i_1, i_2, \ldots, i_k$ with reduced complexity, which it can handle. This would probably increase the binary program code for the software implementation for the processor C_4. For this case, we can also not provide preemption anymore. If we want to migrate a thread from C_4 back to C_3 while processing i_l, where $1 < l < k$, there is no equivalent context for C_3.

Hybrid multi-cores additionally provide reconfigurable hardware cores, where threads are implemented as hardware circuits. As already described in Section 2.3, it is complicated and usually impractical to allow preemption for hardware threads. Instead, most related work use preemption points, i.e. [57, 82, 102]. Lübbers and Platzner have introduced the concept of cooperative multitasking for ReconOS systems focusing on hardware threads in [94]. In this thesis, we extend the concept of cooperative multitasking towards the hardware/software boundary to allow thread migration even between different modalities.

In cooperative multitasking, the threads have well-defined migration points and inform the operating system every time they reach these points. These migration points are defined by the thread designer at the current stage. However, these migration points might be extracted automatically from the data flow graph for streaming applications. For instance, a thread that processes data packets could be migrated in between two data packets. Considering a hardware thread, most parts of the circuit would idle, because the thread waits for the next data packet. In the multithreaded programming model, the thread calls an operating system function to get the next data packet. At this point, the hardware thread is blocked anyway by the operating system and the thread's context is minimal. In between the processing of data packets, the thread context might only consist of application-specific initialization data and additional dynamic parameters. While there is the potential to automatically extract the migration points, this aspect is out of scope for this thesis.

Resuming execution from the migration points should be possible for both, the hardware and the software thread. The thread designer does not only define the migration points but also the thread context manually. At thread migration, the operating system stores the migration point and the corresponding context internally. For a migrated thread, the OS tells the thread, at which migration point it was suspended/migrated. For hardware threads, the OS transfers the context into the thread's local memory. In software, the thread receives a pointer to the memory region, where the context is stored.

Thus, the system can dynamically migrate threads between the hybrid cores. Migrating threads between cores influences various aspects of the hybrid multi-core, such as the application's performance, the overall power-consumption and the on-chip temperature distribution. For instance, migrating a thread from a processor to a hardware core affects the performance, overall power consumption and the thermal profile of the chip. Considering heterogeneous processors, similar effects can be observed for the migration between processors.

The application's performance and the system's power consumption can be influenced by thread migration, because the power consumption of a hardware circuit (thread) with a slow clocking is often significantly less than the power consumption of a processor with fast clocking, which executes the same thread [144]. Similar effects on the performance, can also be observed for heterogeneous processors, which might differ in clocking rates, architecture, or, instruction sets. On current Xilinx FPGAs, we can find different types of processors, such as soft-core processors, i.e. the Xilinx PicoBlaze and the Xilinx MicroBlaze, and hard-core processors, i.e. the PowerPC and the ARM dual-core Cortex-A9. Note that hard-core processors are not part of the reconfigurable FPGA logic. There exist FPGA families that additionally provide these hard-core processors, such as the Xilinx Virtex-II Pro, Virtex-4, Virtex-5 (PowerPC), and Zynq-7000 All Programmable SoC (ARM dual-core).

The heterogeneous/hybrid cores have different temperature characteristics. Thus, migrating threads can not only shift local hot spots from one part of the chip to another part, but can also reduce/eliminate local hot spots (by migrating to a core that has a low temperature profile, such as a processor with slow clocking) or create new local hot spots (by migrating to a core with a high temperature profile, such as a processor with fast clocking). This thesis focuses on the performance and thermal aspects of the threads and only covers the power consumption to some extent, as the self-adaptation algorithms developed for the performance management intend to reduce the number of active cores. When idle cores can be switched into a low-power mode, a reduction of the number of active cores can results in a lower power consumption.

3.4 Goals for Self-adaptation

Self-adaptive systems maintain system goals. These goals can be diverse, e.g., maximizing the throughput or minimizing the mean response time of an application, minimizing the overall power consumption, balancing the on-chip temperature, etc. A self-adaptive system needs the capability to observe how well the system achieves its goals. Whenever a subset of the system goals is not met, the system might need to adapt itself in order to meet its goals in near future.

The trade-off between conflicting goals, such as maximizing an application's throughput and minimizing the overall power consumption, is challenging. One approach to deal with conflicting system goals is to transform the goals into constraints. For instance, the user can define a desired throughput of an application, which forms a system constraint. The system then tries to minimize its power consumption while taking into account the application's performance constraints. Similarly, the system can introduce thermal constraints to combine the goal of balancing the on-chip temperature with an opposing goal such as maximizing an application's performance. It is beneficial to allow the adaptation of the system constraints since it can provide a better adjustment to the system dynamics.

3.4.1 Performance

For performance management, we define streaming applications as target applications in this thesis. Streaming applications are a popular class of applications for embedded systems. Various applications in the domains of multimedia processing, e.g. video and audio processing, and data mining, e.g. pattern recognition, are streaming applications. Many research projects targeting embedded systems focus on this kind of application. Streaming applications have the advantage that the computational blocks can be pipelined and, thus, run in parallel to each other. Often, several instances of the same computational block can be instantiated at the same time such that the workload is distributed to the several instances of the same computational block. Figure 3.4 shows the data flow of a generic streaming application.

The computational blocks can be implemented in software and/or hardware. Parts of an application, which show a high degree of parallelization, can often provide a higher performance when they are implemented in hardware and not in software. At the same time, the sequential parts of an application can be implemented more efficiently using a general-purpose CPU. This composite nature of streaming applications makes them a perfect fit for hybrid multi-core architectures. However, hybrid multi-cores increase the complexity of

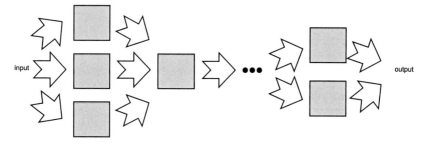

Figure 3.4: Data flow of a generic streaming application: streaming pipeline with one instance (top) and multiple instances (bottom) per computational block.

performance estimation and require efficient and flexible mechanisms for design space exploration.

Dynamically adding and removing instances of the available computational blocks influences the performance (in execution time) of the application. Furthermore, our hybrid multi-core can provide different performance characteristics for a (thread, core) tuple. For instance, a hardware implementation does not provide the same speedup for each computational block compared to its software implementation. Each block can have a different degree of parallelism and is, hence, differently suited for a hardware representation.

We assume that the user defines a performance goal for an application that consists of several computational blocks, which are implemented as hardware and/or software threads. The system measures the application performance and adapts the hardware/software partitioning of the threads at run-time to meet the user-defined performance goals. The system intends to minimize the number of active cores in order to save power. In this sense, the system has additionally the goal to minimize the overall power consumption. In this thesis, we study lower performance bounds and performance intervals for streaming applications.

3.4.2 Temperature

For thermal management, the proposed self-adaptive algorithms aim at balancing the on-chip temperature distribution. To measure the temperature, we divide the chip into a regular grid of tiles, where each tile contains a temperature sensor. The self-adaptation algorithms migrate threads between the hybrid cores in order to minimize the difference between the maximum and the minimum temperature of all tiles for each time step. The temperature is measured discretely at a defined measurement frequency.

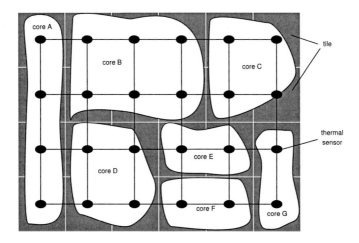

Figure 3.5: Example core map: Seven cores (A–G) are mapped to a 6 × 4 regular grid of tiles where each tile contains a sensor.

Note that the tiles do not represent cores. Due to the hybrid nature of the multi-core, the cores can stretch across several tiles. In this thesis, we assume (i) that each tile can be assigned to at most one core and (ii) that the tiles of each core span a rectangle. An example core map can be seen in Figure 3.5 where seven cores (A–G) are mapped to a 6 × 4 regular grid of tiles. The size of the cores varies from two to six tiles. While the thread-to-core mapping changes at run-time, the core map is fixed.

3.5 Chapter Conclusion

In this chapter, we outlined the multithreading approach for hybrid multi-cores. We presented how a multithreaded application can be mapped to a hybrid multi-core and how different thread-to-core mappings affect several aspects of the hybrid multi-core, such as the application's performance, the on-chip temperature distribution and the overall power consumption. We compared different operating system concepts for multi-core systems and explained, why we use a master-slave approach, which is already applied by ReconOS, in this thesis.

Thread migration (at run-time) is presented as the tool of choice for allowing the hybrid multi-core to react to system dynamics, such as changing requirements or changing workload. Cooperative multitasking is used to allow thread migration between the hardware/software boundary, where the thread designer defines migration points where the thread's context is minimal and processable for both, a hardware and a software implementation. Thus, hybrid multi-cores gain the potential to change the hardware/software partitioning at run-time, which is otherwise fixed. This widens the possibilities for thread-based self-adaptation to achieve effects way beyond state-of-the-art thread migration on homogeneous multi-processor systems. We defined streaming applications as target applications for our self-adaptive hybrid multi-cores.

Furthermore, we formulated the goals for our performance and thermal management. For performance management, the user-defined performance goals for an application can be translated into constraints. This leaves the self-adaptive system the flexibility to select a thread mapping that furthermore achieves the second goal of reducing the power consumption. For thermal management, we divided the chip into a regular grid of tiles, where each tile contains a temperature sensor. Each core can have multiple sensors depending on its size in area. The self-adaptive algorithms minimize the difference between the highest and lowest tile temperature on the chip in order to keep the spatial thermal differences into reasonable bounds.

The next chapter will introduce our hybrid multi-core architecture, the common programming and execution model for software/hardware threads and, finally, the operating system layer that supports run-time adaptation of the thread-to-core mapping.

'Well, perhaps you haven't found it so yet,' said Alice; 'but
when you have to turn into a chrysalis—you will some day,
you know—and then after that into a butterfly, I should
think you'll feel it a little queer, won't you?'
'Not a bit,' said the Caterpillar.

Lewis Carroll, *Alice's Adventures In Wonderland*

CHAPTER 4

Self-adaptive Hybrid Multi-cores

This chapter presents our self-adaptive hybrid multi-core architecture, which
is based on ReconOS [90–96]. The chapter describes the general status of the
ReconOS architecture (including hardware threads, software threads, delegate
threads and the monitoring units) and the ReconOS operating system. Two
ReconOS versions are discussed, which have been used in this thesis. Furthermore,
the applied approaches for software multitasking (on worker CPUs) and hardware
multitasking [94] are discussed. Finally, this chapter presents our methodology
for context-free trans-modal thread migration and gives an outlook for future
developments that include the thread context.

4.1 Architecture

Our self-adaptive hybrid multi-core architecture consists of one main CPU,
multiple worker CPUs and multiple reconfigurable hardware cores. The hardware
architecture for our self-adaptive hybrid multi-core is depicted in Figure 4.1.
The main CPU executes the operating system and possibly software threads of
an application. The operating system consists of a host operating system that
supports multithreading on a single processor, such as eCos or Linux, and several
ReconOS extensions, which are described later in this chapter.

Software threads that run on the worker CPUs and hardware threads that run
on the hardware cores are called **remote threads** and the corresponding cores
are called **worker cores** in the following. Remote threads are represented by

43

delegate software threads on the main CPU. Each time a remote thread calls an operating system service, an interrupt is sent to the main CPU and the call and its parameters are forwarded to the corresponding delegate thread. The delegate thread then calls the operating system (OS) service on behalf of the remote thread and returns the result of the OS call to the corresponding remote thread. It is hereby transparent to the underlying host operating system, if a thread that calls the OS service runs locally on the main CPU, on a worker CPU or even on a hardware core.

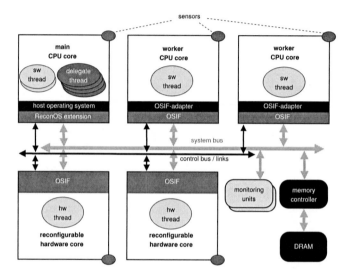

Figure 4.1: Hybrid multi-core architecture overview

Hence, the host operating system does not need to know, if we migrate a thread from one core to another because we hide the thread migration inside our delegate thread. The thread migration will be triggered either explicitly inside an application or, more transparently, inside the ReconOS scheduler, which is part of the ReconOS extension of the host operating system. The ReconOS scheduler is another software thread, which (re)maps the threads to the worker CPUs and the hardware cores.

As can be seen in Figure 4.1, ReconOS uses separate buses for the communication between a remote thread and its delegate thread, on the one side, and for memory access of the worker CPUs and hardware cores, on the other side. This reduces the number of conflicts on the bus compared to a shared bus and significantly lowers the time overhead for calling an OS service on a remote thread. Calling an OS service usually involves only limited data transfer between the calling thread and the master CPU. For instance, if a remote thread wants to lock a semaphore, it has to forward the command and the semaphore identifier to the delegate thread. After locking the semaphore, the answer from the delegate thread is a signal that the semaphore is successfully locked. While the delegate thread waits for the semaphore, there is no communication between the thread and its delegate. In contrast, memory accesses might involve burst transfers to/from the main memory (DRAM[21]), which fully utilize the bus for a longer time period.

On a worker core both communication types are handled by the operating system interface (OSIF). Since a processor has different communication ports compared to a ReconOS-based hardware core, an OSIF-adapter translates the communication ports and signals of a processor to the ports and signals that are expected by the OSIF.

Finally, the architecture includes core-specific sensors. For temperature measurements, we have divided the chip into a regular grid of tiles, where each tile contains a sensor, as defined in Section 3.4.2. The spatial core map is fixed for the multi-core. Each core covers at least a single tile of the chip. In our system, we associate all temperature sensors to the core that are covered by it. Hence, each core can have multiple temperature sensors. In our current implementation the temperature readings are performed globally on the master processor, but in future implementation these readings might also be processed on each core individually by the OSIFs. For performance measurements, we use timers, which can measure the execution time of a thread.

For a host operating system that uses a virtual address space such as Linux, the MMU has to translate virtual addresses to physical addresses. To provide an efficient address translation, the MMU contains a shared translation look-aside buffer (TLB), which caches address translations. The MMU can autonomously perform a page table walk through the Linux kernel's page tables to handle TLB misses. Finally, the MMU can provide access control, where an unauthorized memory access returns an error code to the corresponding worker core. More information about the hardware memory virtualization in ReconOS can be found in [21].

---

[21] dynamic random-access memory (DRAM)

ReconOS version 2.01

In ReconOS version 2.01, a hard-core PowerPC is used as main CPU and a second hard-core PowerPC can be optionally used as worker CPU. The worker CPU can be clocked at a different frequency than the main CPU to create heterogeneity between the processors. The OSIFs of the hardware threads use the processor local bus (PLB) as system bus to access the main memory. In this version each OSIF has an individual master bus attachment to the PLB. Since the number of bus masters is limited for the used PLB, the maximum number of hardware threads is limited to seven in version 2.01. The operating system interface supports single word read and write operations and also burst transfers up to a length of 32 words, where a word contains four bytes.

Two host operating systems are supported in version 2.01: the rich featured operating system Linux and the embedded operating system eCos. For Linux-based ReconOS designs, virtual memory is used. Therefore, each OSIF contains an individual memory management unit. ReconOS v. 2.01 supports dynamic partial reconfiguration of hardware threads at run-time and cooperative hardware multitasking.

ReconOS version 3.0

Figure 4.2 shows the latest architectural overview of a ReconOS system with the version 3.0. A soft-core MicroBlaze processor was used as the main CPU instead of a PowerPC. We switched to a MicroBlaze processor since the hard-core PowerPC is not anymore integrated in the latest FPGA families of Xilinx.

Using soft-core CPUs instead of hard-core CPUs has the additional advantage that we can instantiate an arbitrary amount of such processors on any FPGA as long as there is enough resources to do so. Furthermore, we can configure the soft-core processors individually. For instance, CPUs can be configured with different cache sizes, multiple variants of hardware accelerators (such as floating point units) and various clock frequencies. Note that different clock frequencies can also be applied to the PowerPCs on ReconOS v2.01 systems. But the potential for configuration was limited as the circuit of the PowerPC could not be changed on the FPGA.

Porting the main CPU to a MicroBlaze CPU required further modifications in ReconOS. ReconOS v3.0 uses two fast simplex links (FSLs) for bidirectional communication between each OSIF and the main CPU. The MicroBlaze supports up to 32 FSL ports. ReconOS v3.0 supports up to 14 remote cores, because it needs the remaining four FSL ports for the ReconOS control system.

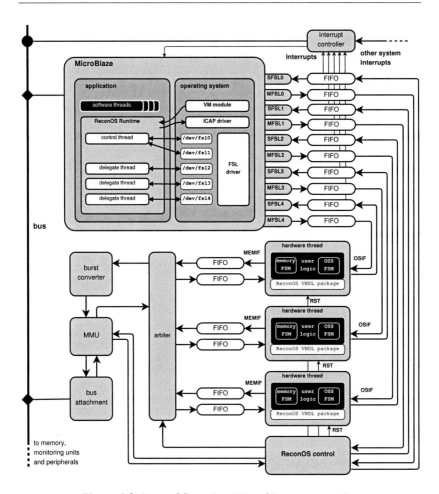

Figure 4.2: ReconOS version 3.0 architecture overview

The ReconOS control system has been introduced in ReconOS v3.0 to handle the memory accesses of the worker cores in a shared controller. This reduces the resource overhead of the OSIFs of the worker cores. In ReconOS v3.0, there exists a central memory management unit (MMU), which reduces the number of required PLB master attachments to one for all worker cores. This means that ReconOS is no longer restricted by the maximum number of supported PLB bus

47

masters. Hence, the maximal number of supported worker cores could doubled from seven (version 2.01) to 14 (version 3.0).

The memory accesses are handled by an arbiter, which is connected to all worker cores using FIFO links via the memory interface (MEMIF). To allow greater efficiency for burst accesses to the memory, the buffer sizes for the FIFO links are increased compared to the buffer sizes of the FSLs to the main CPU, which only have to store the information for one OS call. A shared burst converter splits up memory bursts that exceed the buffer sizes of the FIFO links to allow burst transfers of arbitrary length. This arbiter has been introduced first in ReconOS v3.0. An MMU is attached to a bus master and performs the memory accesses. In ReconOS v. 3.0 all hardware threads share a single MMU. In this version Linux is again supported as host operating system. However, eCos has been replaced by the Xilkernel [163] of Xilinx, which is a thin library that provides rudimentary operating system functionalities.

The latests version of ReconOS has not the entire functionality of ReconOS v. 2.01 to this point in time. Worker processors, dynamic partial reconfiguration and cooperative hardware multitasking are not supported yet.

4.2 Software Threads

On our hybrid multi-core, threads can be implemented in hardware and/or software. In the following, we illustrate the programming of a software thread and a hardware thread for a sorting case study, see Figure 4.3. In the sorting case study, a thread, which is either implemented in software or in hardware, sorts data blocks of fixed size. The data is stored in main memory and the thread receives pointers to unsorted data blocks as messages in the incoming message box mb_unsorted. After sorting the data block, the threads forwards the same pointer to the outgoing message box mb_sorted. The data block size is fixed to a predefined size N for this case study.

Source code 4.1 shows the software implementation of the sorting thread in the programming language C. The incoming message is read first (line 5), before the data block at address ptr is sorted by the bubble sort algorithm, line 6. Finally, the message is forwarded to the outgoing message box to indicate that the data block is now sorted, line 7.

This software implementation can be used for the main processor and also for the worker processors. On the worker processors, the thread designer has to include the library kapi_cpuhwt.h. This library implements the kernel API for the worker CPU. As there is no operating system on the worker CPU, the remote

```
1  void* sorting_thread( void * arg ) {
2     void *ptr;
3
4     while ( true ) {
5         ptr = mbox_get(mb_unsorted);
6         bubblesort( (unsigned int*) ptr, N);
7         mbox_put(mb_sorted, ptr);
8     }
9  }
```

Source code 4.1: C source code of a software thread that sorts a data block

software thread is the only thread that is executed on the processor. Therefore, a main() function just has to call the thread, i.e. the sorting thread.

When the software thread calls an OS function (e.g. mbox_get), the thread is blocked and the command together with its parameters is written into the input registers of the OSIF-adapter. A command decoder then forwards the request via the OSIF to the corresponding delegate thread and waits for the OS call to complete. The return values of the OS call are written into the output registers of the adapter, which are read by the worker processor before the OS function gets unblocked. Since only one thread is executed on the worker processor, the entire processor gets blocked while performing an OS call. For worker CPUs, the software implementation is compiled into an executable file.

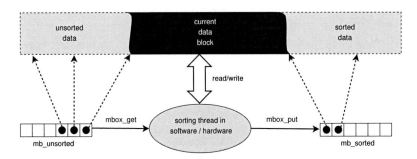

Figure 4.3: Sorting case study: A sorting thread receives a pointer from the message box mb_unsorted that points to an unsorted data block in the main memory. After sorting, the thread puts a message into the outgoing message box mb_sorted.

The described approach provides an abstraction allowing cores with different instruction set architectures to be transparently integrated into ReconOS. Other threads in the system as well as large parts of the OS kernel itself do not need to know on what kind of a core (software or hardware) a particular thread runs. This is because threads that are not executed directly on the master processor and thus are not under direct control of the regular OS scheduler are nevertheless represented on the master CPU through their delegate threads.

Before the system begins the execution of the software thread on the worker CPU, the delegate thread has to be created on the main CPU first. After the delegate thread is created, a reset signal is sent to the worker CPU. A reset logic block inside the OSIF-adapter prevents the booting of the worker processor before a delegate thread is started. The booting code is stored into a block memory (BRAM) on the FPGA. The booting code includes the jump address to the executable in the main memory.

4.3 Hardware Threads

A hardware thread in ReconOS is divided into two parts: The first part is the user logic that contains the highly parallel computational logic of the thread or, in other words, the actual functionality of the thread. The second part contains one or two finite-state machines that control the user logic, perform operating system calls, and access the shared memory. A finite-state machine that handles operating system services is called OSS FSM and a finite-state machine that handles the shared memory access is called memory FSM, as depicted in Figure 4.2. Both FSMs can alternatively be combined to a single finite-state machine called ReconOS FSM.

A hardware thread includes local memory to allow efficient computation. First, a hardware thread can buffer the required data into its local memory using efficient burst reads from the main memory. Then, it can process the data locally by exploiting the direct access and low latency of the BRAM. Finally, the thread can write back its altered data to the main memory using efficient write bursts. The local memory can be seen as a local data cache or buffer for the hardware threads with extreme low time overhead and without any conflicts caused by other threads in the system.

To ease the programming of hardware threads, ReconOS provides a VHDL library that contains a POSIX-like API to access operating system services and, furthermore, read/write access to the main memory of arbitrary length. Whenever a finite-state machine calls an OS service or performs a read/write access to the shared memory, the corresponding state machine is blocked. When an OS service is called, the OSIF forwards the call to the corresponding delegate on the master processor. The delegate calls the OS service on behalf of the hardware thread and returns the result of the OS call back to the hardware thread. When the OSIF receives the result, it unblocks the finite-state machine. The VHDL procedure then returns the result of the OS call. For memory accesses, the finite-state machine is blocked until the data is fully transfered between the local memory and the main memory.

Source code 4.2 shows the VHDL code for the sorting thread where a ReconOS FSM handles both, the OS service calls, and, the memory accesses of the hardware thread. First, the FSM starts in the state STATE_GET_ADDR where the thread waits for a message from the message box MB_UNSORTED by calling the function osif_mbox_get (lines 17–21). The message contains a pointer ptr to an unsorted data block. In the next state, STATE_READ, the hardware thread copies the entire data block to the local memory at the address addr_0 using the function memif_read (lines 23–28). The signal len describes the length of the data block in bytes.

After transferring the data block to the local memory, the state machine activates the user logic by setting the sort_start to '1' where the user logic sorts the data using the bubble sort algorithm in the state STATE_SORTING (lines 30-34). The VHDL code for the bubble sort algorithm is omitted for simplicity. The user logic signals the end of the sorting using the sort_done signal. When this signal is set, the state machine goes to the next state called STATE_WRITE where the data block is transfered back to the main memory (lines 36-40). Finally, the FSM sends the message to the message box MB_SORTED that contains the pointers to the sorted data blocks (lines 42–45).

Compared to Source code 4.1 of the software implementation, additional read and write memory accesses are required. Apart from that, the FSM of the hardware thread can be programmed in the same manner as the software thread. The additional parameter for the OS service calls and the shared memory accesses, such as i/o_osif, i/o_memif and i/o_ram indicate the input and output ports of the interfaces to the OSIF, the MEMIF and the local RAM of the hardware thread. These interface have to be reset initially (lines 6–8). The done variable indicates if a blocking call via the OSIF or the MEMIF is finished. Finally, the result signal, line 43, forwards the delegate's result of the OS call to the hardware thread.

A hardware core contains an operating system interface, a memory interface and a rectangular-shaped partial reconfigurable area. Different hardware threads can be reconfigured into the same hardware core. The entire FPGA design of the hybrid multi-core can be divided into a static part and a dynamic part. The static part contains most parts of the system, such as the processors, the memory

```vhdl
 1  -- operating system services and memory finite-state machine
 2  reconos_fsm : process (clk, rst) is
 3      variable done : boolean;
 4  begin
 5      if rst = '1' then
 6          osif_reset(o_osif);
 7          memif_reset(o_memif);
 8          ram_reset(o_ram);
 9          done := False;
10          len <= conv_std_logic_vector(N,24);
11          addr_0 <= X"00000000";
12          sort_start <= '0';
13          state <= STATE_GET_ADDR;
14      elsif rising_edge(clk) then
15          case state is
16
17              when STATE_GET_ADDR =>
18                  osif_mbox_get(i_osif, o_osif, MB_UNSORTED, ptr, done);
19                  if done then
20                      state <= STATE_READ;
21                  end if;
22
23              when STATE_READ =>
24                  memif_read(i_ram, o_ram, i_memif, o_memif, ptr, addr_0,
25                             len, done);
26                  if done then
27                      sort_start <= '1';
28                      state <= STATE_SORTING;
29                  end if;
30
31              when STATE_SORTING =>
32                  sort_start <= '0';
33                  if sort_done = '1' then
34                      state <= STATE_WRITE;
35                  end if;
36
37              when STATE_WRITE =>
38                  memif_write(i_ram, o_ram, i_memif, o_memif, addr_0, ptr,
39                              len, done);
40                  if done then
41                      state <= STATE_ACK;
42                  end if;
43
44              when STATE_ACK =>
45                  osif_mbox_put(i_osif, o_osif, MB_SORTED, ptr, result, done);
46                  if done then
47                      state <= STATE_GET_ADDR;
48                  end if;
49          end case;
50      end if;
51  end process;
```

Source code 4.2: VHDL code of a hardware thread that sorts a data block controller, the system bus, the OSIF and the MEMIF of each hardware core, etc. In contrast the dynamic part contains only the reconfigurable regions of the hardware cores. Each hardware thread is implemented in such a way that it is connected to the interfaces of the corresponding hardware core using the same bus macros.

The reconfiguration is triggered and handled by a scheduler that runs on the master processor. More details about a thread migration between different cores and multithreading on one core will be given in Section 4.9.

4.4 Delegate Threads

When a remote thread calls an OS service it is blocked until its delegate returns the answer. At an OS call, the OSIF first sends an interrupt to the main CPU. Each worker core has its individual interrupt. Since there is at maximum one active remote thread being executed on each worker core, an interrupt handler can wake up the corresponding delegate thread. Hence, the interrupt can be directly assigned to the corresponding delegate thread by the interrupt handler. Next, the corresponding delegate thread reads the OS call command and its parameter via the control bus or the control links. The delegate thread then calls the OS call on behalf of the remote thread, returns the result to the calling remote thread using the control bus/links and finally unblocks the remote thread.

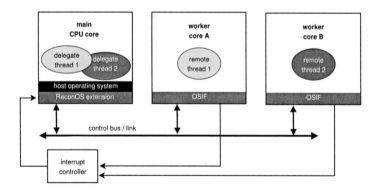

Figure 4.4: Delegate thread overview for a multi-core with two worker cores.

Figure 4.4 shows an example for two worker cores *A* and *B*. The delegate threads were initially developed to integrate remote hardware threads into the system, but for hybrid multi-cores the same approach is used for remote software threads being executed on worker processors.

Note that it is also possible to generate static hardware cores, too. These cores differ from reconfigurable hardware cores in that they can not be reconfigured at run-time and, thus, always execute the same hardware thread. However, for our self-adaptive hybrid multi-core we consider reconfigurable hardware cores.

Section 4.7 presents the applied software multitasking approach, which allows multiple software threads being executed on the same worker processor, and Section 4.8 describes the hardware multitasking approach to allow multiple hardware threads being executed on the same hardware core. Each remote thread is hereby represented by a delegate thread. In contrast to a reconfigurable hardware core, no FPGA resources have to be reconfigured, when the ReconOS scheduler or the application decides to map a different thread on a worker processor.

4.5 Operating System Layer

ReconOS extends monolithic operating system kernels that support multithreading for single processors. In our self-adaptive hybrid multi-cores, the extended host operating system is executed on the master processor. Threads being executed on worker cores, such as worker processors and reconfigurable hardware cores, are represented by delegate threads that run on the main processor. The ReconOS extension for any operating system, such as Linux or eCos, supports these delegates. ReconOS provides an API, which an application designer can use to handle the delegates. For instance, there are dedicated functions for delegate creation and termination.

When a remote thread is started this involves the creation of the delegate software thread on the main processor, the creation of certain data structures for the remote threads (i.e. the list of the OS resources that can be accessed by the thread) and either the reconfiguration of a hardware core (for hardware threads) or the reset of the worker processor (for software threads). Furthermore, the interrupt of the corresponding worker core has to be associated with the delegate thread. The delegate threads run at a high priority because their only task (besides starting the thread on the worker core) is the handling of OS calls. We assume that our remote threads perform OS calls only infrequently. Thus, the delegates only utilize a small amount of the available execution time of the main processor. Handling single OS calls requires only a low overhead in time, hence, a high priority is justifiable.

A ReconOS scheduler, which is implemented as a software thread, manages the thread-to-core mapping. The scheduler implements the self-adaptive strategies of the multi-core. This means that the scheduler also triggers the thread migrations between heterogeneous or hybrid cores.

The operating systems eCos and Linux have been extended to support the ReconOS programming and execution model. We target embedded real-time applications for eCos-based systems. The embedded operating system eCos is entirely composed by packages and is therefore highly configurable. Furthermore, it provides a small memory footprint and has a reduced complexity compared to rich-featured operating systems, such as Linux. However, there is no memory protection in eCos. In contrast, Linux is a Unix-like operating system that supports memory protection, differentiates between kernel space and user space, and contains a wide set of features and drivers. However, compared to eCos, Linux provides a high overhead in memory usage and time, i.e. for task swapping and OS call handling. Since ReconOS version 3.0, eCos has been replaced by the Xilinx Xilkernel. More details on the operating system extensions and the tool flow can be found in [90].

4.6 Monitoring

For our self-adaptive hybrid multi-core, we employ a global performance sensor and a regular grid of temperature sensors. Both sensor types will be discussed in this section.

Performance Sensor

We use a shared time base to measure the execution time of a thread or an entire application. The time base is connected to the system bus and contains a register. The register value is initialized to 0 when the system is reset, and is increased by one for each clock cycle. To measure the execution time, the time base has to be read at the start of the measurement, v_{start}, and at the end of the measurement, v_{end}. Then, the measured time can be computed as follows:

$$t = \frac{v_{end} - v_{start}}{f} \tag{4.1}$$

where t is the measured execution time and f is the clock frequency (in Hz) of the time base.

If the end value of the measurement, v_{end}, is lower than the start value, v_{start}, there was (at least) one overflow for the register value. As the time base increases its value continuously an overflow happens at least each 42.94 seconds for a time base that contains a 32-bit register and that is clocked at 100 MHz. For this case the maximum register value r_{max} has to be added:

$$t = \frac{r_{max} + v_{end} - v_{start}}{f} \qquad (4.2)$$

The maximum measurement period is limited to 42.93 seconds for this scenario. However, longer measurement periods can be achieved by extending the register to 64 bits where a time base clocked at 100 MHz can run for centuries without any overflow. The sensor layout is depicted in Figure 4.5.

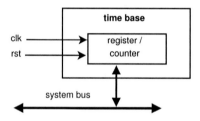

Figure 4.5: Performance sensor layout: A register is increased each clock cycle and its value can be read using the system bus.

Temperature Sensor

Similar to related work [108, 148, 170], we employ ring oscillators as temperature sensors on our FPGA-based systems. A basic ring oscillator is a simple circuit composed of an odd number of inverters as shown in Figure 4.6. Its output Q oscillates between 0 and 1 at a frequency f based on the delay of the inverters. The delay again depends on the current temperature, more precisely a high temperature leads to a long delay and thus to a low frequency f. This dependency makes it possible to use ring oscillators as temperature sensors by measuring their frequencies.

To design a robust temperature sensor we modify the basic ring oscillator as shown in Figure 4.6. An AND-Gate in the chain of inverters is used to connect an enabling signal to the sensor. When no measurement is done, the oscillator is disabled, which reduces its power consumption and its thermal impact on the system. Furthermore, a flip-flop takes the output Q of the ring oscillator as its clock signal. The flip-flop updates its value on a rising edge in order to provide a clean output-signal Q'. The sensors increases a counter every time Q' changes its value from 0 to 1. We use the counter value to determine the frequency of the ring oscillator. Note that the sampling-frequency has to be at least twice as high as the frequency of the ring oscillator to provide reliable measurements.

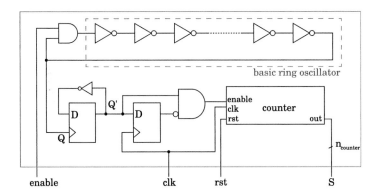

Figure 4.6: Temperature sensor layout: A ring oscillator with an odd number of inverters is used for temperature measurements. [12]

A temperature measurement includes the execution of following steps:

1. Enable the ring oscillator

2. Wait x_1 clock cycles so that the ring oscillator can gain a stable frequency

3. Sample the output Q' for x_2 clock cycles and output the counter value

4. Disable the ring oscillator

As described in Section 3.4.2, we divide the FPGA into a regular grid of tiles where each tile contains a temperature sensor. A central monitoring unit controls all temperature sensors. The monitoring unit is attached to the system bus. The unit can be controlled by a monitoring thread that runs on the main processor as part of the ReconOS extension. In contrast to related work, we have developed a novel self-calibration technique, which is presented in Section 6.1.

4.7 Software Multitasking

Multitasking is known as a technique that allows multiple tasks, i.e. threads, to run at the 'same' time. On CPUs multitasking is usually provided using preemption. At preemptive multitasking, the active task can be suspended from the core and replaced by another task at arbitrary points in time. A scheduler implements an algorithm, such as time slicing, that defines the time schedule for all tasks. When the active task is replaced by another task, this also implicates a context switch. For small time slices, it appears to the user that multiple tasks (or threads) are executed at the same time.

On the main CPU, preemptive multitasking is supported out-of-the-box by the host operating system. However, we do not support preemptive multitasking on the worker CPUs to maintain the simplicity of the system design. In our current implementation we only support one software thread at a time on each worker CPU. There are multiple possible approaches to replace the current software thread on a worker CPU with a different software thread:

Use multiple executables, copy the new executable to the jump address of the boot code and reboot the worker CPU. This would come with an overhead of copying the new executable to the jump address. While the old executable is overwritten, the system needs to ensure that the worker CPU is stalled to avoid an unforeseeable behavior in program execution. To replace a software thread with another one, the delegate thread of the current remote software thread has to be terminated first, then the worker CPU has to be stalled, the next executable has to be copied to the jump address and finally the new delegate thread has to be created.

Use multiple executables and reboot at another jump address. The jump address in the boot code can be altered to the memory location of the new executable. This would save the time and memory overhead to copy the executable to the jump address.

Combine multiple software threads into a single executable and select the current thread using an identifier internally after a reboot. When a software thread is instantiated on a worker CPU, a thread identifier is additionally provided as initial data. The OSIF contains a data register that can store an initial data value. This register could be used to store the thread identifier of the remote thread.

In the `main()` function of the executable on the worker processor, the data register that contains the current thread identifier is read first, before the corresponding thread is started. However, to switch between threads A and B, the delegate thread of A must be terminated first, before a new delegate can be created for B, which also results in a reboot of the worker CPU. After rebooting, the `main()` function is called again, which then selects the software thread B. A hand-coded example for this approach can be seen in Source code 4.3 where three different software threads can be started depending on the thread identifier `thread_id`, which can be accessed using the function `getInitData()`.

Using this approach, the boot code of the system does not need to be modified at run-time. Additionally, this reduces the number of executables per worker CPU to one. For these reasons, we have used this approach in our self-adaptive hybrid multi-core as proof-of-concept for our experiments. However, in future we plan to investigate techniques to support preemptive multitasking on worker CPUs such as the following approach.

Introduce a reduced operating system on each worker CPU , which includes a scheduler and is able to swap threads, in order to introduce preemptive multitasking on worker CPUs. The OS calls however should be still handled by the delegate threads on the main CPU. Here, every software thread that runs on a worker CPU is still represented by a delegate thread. To direct the OS call of a remote thread to the corresponding delegate, there needs to be additional information. In the current ReconOS system, each worker core has its individual interrupt. Hence, if the main CPU receives an interrupt, it can instantly forward the OS call to the corresponding delegate thread, because there is only one active delegate thread per core. Using this approach, the interrupt handler must additionally receive a thread identifier to still be able to forward the OS call to the corresponding delegate. This approach shows a great flexibility, provides a low overhead for multitasking, and avoids a reboot of the CPU. However, this approach requires a high implementation effort and, thus, has not been introduced to ReconOS yet.

4.8 Hardware Multitasking

Multiple hardware threads can be performed on a single reconfigurable hardware cores in three manners, using non-preemptive, preemptive, or, cooperative multi-tasking. In non-preemptive systems, a thread can not be preempted at all time and, thus, runs from start to completion uninterrupted, which minimizes the flexibility of a system to react to any dynamic changes. Therefore, non-preemptive multitasking is not desired for self-adaptive hybrid multi-cores. In contrast to non-preemptive systems, preemptive systems can suspend a running thread T_1 at any point in time, execute a different thread in between and then resume T_1

```
 1          int main(void) {
 2
 3              int * thread_id = (int *) getInitData();
 4
 5              switch(*thread_id){
 6
 7                  case 0:
 8                      thread_a(0);
 9                      break;
10
11                  case 1:
12                      thread_b(0);
13                      break;
14
15                  case 2:
16                      thread_c(0);
17                      break;
18
19                  default:
20                      break;
21              }
22          }
```

Source code 4.3: main() function that can select between three software threads

at a later point in time. However, implementing preemptive multitasking for hardware threads generates an immense overhead in FPGA logic to ensure that a hardware thread can be preempted at any time [76, 131].

ReconOS applies a cooperative approach to reduce this overhead. In cooperative multitasking, the operating system and the hardware thread work together in order to find adequate preemption points, where the thread context is well-defined and minimal, and in order to transfer the thread context. By applying cooperative multitasking, the overhead in FPGA logic is drastically reduced compared to a multi-core applying preemptive multitasking, since the context handling has been simplified. Cooperative multitasking can be seen as a compromise between non-preemptive and preemptive multitasking allowing infrequent thread preemptions at the cost of an acceptable hardware overhead.

Unlike the context of a software thread, the context of a hardware thread is not well-defined. The reconfigurable region of the hardware core contains a vast amount of storage elements. It is not intuitive to know which storage elements are currently used when a hardware thread is preempted. Furthermore, different hardware threads utilize different subsets of storage elements. Thus, for an efficient context switch that does not save the values of all available storage elements, extra logic would be needed. For instance, scan chains could be integrated into each circuit of a hardware thread that allows the operating system to extract the state context. However, including such scan chains into the highly parallel user logic of a hardware thread creates a non-negligible hardware overhead.

Instead, for cooperative multitasking, the programmer manually defines suitable yielding points where the hardware thread has a minimal context. Hence, we believe that cooperative multitasking is a good trade-off between the interruptibility of a hardware thread and the corresponding costs in FPGA logic and reconfiguration time (possibly including the context switch). In ReconOS, the yielding points are usually combined with OS calls where the hardware thread is blocked anyway.

When a hardware thread A reaches a yielding point, it informs the operating system that it can be preempted and replaced by another hardware thread. If the hardware thread A has a context besides the current state of its final state machine, this context has to be stored into the local RAM of the hardware thread. If the OS wants to replace the hardware thread A with another hardware thread B, the OS stores the thread context of A and the current yielding point (i.e. the current state of the ReconOS FSM) into the main memory. Then, the OSIF and MEMIF (ReconOS v3.0) are reset and, therefore, deactivated, before the partial bitstream of the hardware thread B is reconfigured to the hardware core.

If the hardware thread B has been preempted before as well, the OS transfers its context into the local memory and informs the hardware thread B at which yielding point it was suspended from execution. Finally, the OSIF and MEMIF are activated again and the hardware thread B resumes its execution.

Source code 4.4 shows how a yielding point is integrated into the sorting hardware thread, which has been presented before in Source code 4.2. The hardware thread yields each time, when it waits for an incoming message, line 27. The OSIF forwards not only the command for the osif_mbox_get call but sets additionally a flag that the thread can be preempted by the OS. The ReconOS scheduler then decides if it wants to replace the hardware thread. The thread stores its context beforehand because it does not know in advance if it will be suspended from execution. In this example, there is no thread context and, hence, no context needs to be stored. Nevertheless, in Source code 4.4 there is a place holder for the context saving (lines 21–23). The state STATE_GET_ADDR is split into two states to include the context saving into the state machine without having to update the rest of the source code. Note that the ReconOS FSM in Source code 4.4 has been shortened for clarity.

```
 1  -- operating  system  services  and  memory  final  state  machine
 2  reconos_fsm :  process  ( clk , rst )  is
 3      variable  done , preempt  :  boolean ;
 4  begin
 5      if  rst  =  '1 '  then
 6          [ ... ]
 7          state  <=  STATE_RESUME;
 8      elsif  rising_edge ( clk )  then
 9          case  state  is
10
11              when  STATE_RESUME  =>
12                  osif_thread_resume ( i_osif , o_osif , resume , preempt , done );
13                  if  done  then
14                      if  preempt  and  resume  =  X" 00000001"  then
15                          state  <=  STATE_GET_ADDR_2;
16                      else
17                          state  <=  STATE_GET_ADDR;
18                      end  if ;
19                  end  if ;
20
21              when  STATE_GET_ADDR  =>
22                  -- store  thread  context :  here ,  there  is  no  context
23                  state  <=  STATE_GET_ADDR_2;
24
25              when  STATE_GET_ADDR_2  =>
26                  -- yielding  point
27                  osif_flag_yield ( i_osif , o_osif , X" 00000001" );
28                  osif_mbox_get ( i_osif , o_osif , MB_UNSORTED, ptr , done );
29                  if  done  then
30                      if  preempt  then
31                          preempt  :=  False ;
32                          state  <=  STATE_LOAD_CONTEXT;
33                      else
34                          state  <=  STATE_READ;
35                      end  if ;
36                  end  if ;
37
38              when  STATE_LOAD_CONTEXT  =>
39                  -- this  thread  has  no  context
40                  state  <=  STATE_READ;
41
42              when  STATE_READ  =>
43                  [ ... ]
44          end  case ;
45      end  if ;
46  end  process ;
```

Source code 4.4: VHDL source code of a cooperative hardware thread

A hardware thread does not know after its configuration if it was preempted before. Thus, the thread initially checks if it was preempted before by calling osif_thread_resume. If the hardware thread was preempted before, the function returns this information along with a binary code that either encodes the state of the FSM or the yielding point. Here, the yielding point was encoded by X"00000001", see line 27. Thus, when the thread resumes execution, it directly switches its state to STATE_GET_ADDR_2 in contrast to STATE_GET_ADDR, lines 14–18.

When the thread has been preempted inside an OS call, the OS command is not sent again to the OS via the OSIF because the delegate has already received the call before the thread was suspended. Instead, the thread only waits for the delegate to finish the still pending OS call. If the sorting thread was preempted before, it has to restore its context before it further processes its execution, lines 30–35. Note that a hardware thread, which is currently not configured into any hardware core, should not replace an already configured hardware thread, if it was directly blocked again because its delegate still waits for an OS resource, such as a semaphore. If the hardware multitasking is managed by the ReconOS scheduler, this situation does not occur, since the scheduler only reconfigures threads that are currently not blocked.

As the reconfiguration of a hardware core takes a significant amount of time (e.g. several milliseconds for a reconfiguration of a bitstream), the scheduler should not swap threads as often as possible. But if a hardware thread is rarely preempted the overhead for context saving, which is performed each time the thread reaches a yielding point, imposes an unnecessary time overhead. Therefore, the operating system can also send the information when it intends to reconfigure a hardware core to the currently active hardware thread. Then, the hardware thread can first check, if there is a yielding request by the OS before it stores its thread context and allows the OS to suspend the thread. A hardware thread can check if there is a yielding call by calling the osif_check_yield function. Since the sorting thread has no context, this option is not integrated into Source code 4.4. Cooperative hardware multitasking was already implemented inside ReconOS in a previous thesis by Enno Lübbers [90]. This thesis conceptually extends the cooperative multitasking approach to hybrid threads that can execute in hardware and in software, which is described in the next section. More details on hardware multitasking in ReconOS can be found in [94].

4.9 Trans-modal Thread Migration

Trans-modal thread migration between CPUs and hardware cores impose many challenges. The toughest challenge is the translation of the context from software to hardware, or vice versa. As described in Section 3.3, one possibility to allow trans-modal thread migration is the extension of the cooperative multitasking approach from hardware threads to hybrid threads, which can be executed in hardware and in software.

Worker CPUs are connected to the main CPU in the same manner as the hardware cores are connected to it. For both core types, the OS calls are forwarded via an OSIF to the corresponding delegates. Hence, the cooperative methods that are already available for hardware threads could also be implemented for remote software threads. If the hardware and software implementations then integrate common migration points at the same positions in execution, context-free trans-modal thread migrations can be supported. Similar to hardware threads, software threads can internally use a final state machine to define migration points. When a software thread is configured to a worker CPU it has to verify, if it was preempted before. If this is the case, the migration point has to be read using the OSIF. This already works for hardware threads. Thus, the OSIF library for remote software threads only needs to be extended to support thread yields and thread resumes as well.

Handling trans-modal thread migrations including the thread context is challenging. Here, it has to be ensured that the state context can be processed by the worker CPUs and the hardware cores. Because we only need to handle manually selected migration points with minimal context, a common context can possibly be found for both modalities. For instance, in a video-based streaming application, the context in-between frames might only include a certain set of learned parameters. Similar to hardware multitasking, a remote software thread might store its context into either a local memory or into the main memory. The OS is also in charge to handle the context migrations. The mechanisms to store and load a thread context can be handled similarly for software and hardware threads.

However, integrating migration points into software threads poses an additional burden for software developers. For software-based systems, the developer usually does not need to take care about thread preemption because the OS is able to handle task preemption transparently. However, for trans-modal thread migrations, migration points are needed that are processable in both, hardware and software. If the applications are given as synchronous data flow graphs or Kahn networks, suitable migration points might be extracted automatically. Thus, future tool support for integrating such migration points for both modalities appears feasible.

In our experiments, trans-modal thread migrations were emulated using dynamic thread creations and terminations. This is possible because this thesis focuses on periodically context-free multithreaded streaming applications. Every time when a thread waits for new incoming data, the thread might be terminated on its core and recreated on another core. This approach emulates context-free (trans-modal) thread migrations without the need to integrate migration points in hardware or in software. Using the emulation approach, multiple instances of the same thread can be created on different cores, which extends the question, on which core a thread should be executed, to the more advanced questions of how many thread instances are needed and on which cores should they run. Finally, this emulation approach also allows for (trans-modal) thread migration from/to the main CPU. Mixing preemptive and cooperative multitasking on the host operating system would implicate deep changes in the OS kernel, i.e. the internal scheduler and the dispatcher. This stands in conflict with the current ReconOS approach that intends to abstain from deep kernel modifications.

4.10 Contributions to ReconOS

Several researchers have contributed to ReconOS over the last years. The first two versions of ReconOS have been developed as part of the dissertation of Enno Lübbers [90]. This thesis has provided a real-world case study, a video object tracker, for the second ReconOS version. The video object tracker was used to test several ReconOS features, such as cooperative hardware multitasking and the integration of worker processors, in a realistic scenario. The proposed method for software multitasking on worker processors, see Section 4.7, and the proposed temperature sensor, see Section 4.6, have been designed, implemented, and tested as part of this thesis.

This thesis has made the following contributions to ReconOS version 3.0:

Operating system interface: We have implemented and validated several operating system calls, e.g. semaphores and conditional variables, as part of the OSIF. Furthermore, we have developed ReconOS queues, which can transfer messages of arbitrary length between different threads. ReconOS queues have been introduced to version 3.0 because the used Linux version did not support message queues (mq) on the MicroBlaze processor. Hence, we have programmed a user-space alternative, which can transfer the messages via the OSIF to/from hardware threads.

Memory interface: We have implemented the API functions for read and write accesses to the shared memory as part of the memory interface, such that transfers of arbitrary length are supported inside hardware threads.

Reference designs: Several reference designs have been developed as part of this thesis, such as reference designs that include temperature sensors and reference designs that support caching for the master CPU.

Case studies: The video object tracker has been ported to the latest ReconOS version. Furthermore, a second webcam case study has been designed and implemented. The webcam case study receives video frames from a webcam, which is connected to a workstation, using the Ethernet interface. The application then manipulates the frames using hardware and/or software threads and sends them back to the workstation. The original and the manipulated frame are displayed next to each other in a graphical user interface on the workstation. This case study has been used in several tutorials to give new ReconOS users an opportunity to program their own threads in hardware and software and to verify the result of their implementations using a graphical output.

4.11 Chapter Conclusion

This chapter established the hybrid multi-core architecture and the common programming and execution model for software and hardware multithreading, which is based on ReconOS. A main processor executes the entire operating system. The worker cores are connected to the main processor using either a control bus or FSL links (depending on the ReconOS version). The monitoring units and the main memory are connected to all cores using a shared PLB bus. For embedded applications, ReconOS extends the embedded operating system eCos (ReconOS v. 2.01) or the Xilinx Xilkernel (ReconOS v. 3.0), and, for high-performance applications, ReconOS extends the rich-featured operating system Linux. For the purpose of monitoring the system state, the multi-core architecture includes a time base as performance sensor and a regular grid of ring oscillators as temperature sensors.

Furthermore, this chapter presented the employed approaches for software multitasking on worker processors and hardware multitasking on reconfigurable hardware cores. A cooperative approach is applied for hardware multitasking. If software threads are programmed in the same cooperative manner, this allows for thread migration even across the hardware/software boundary. In this thesis, (trans-modal) thread migrations are emulated by dynamic thread creations and thread terminations.

The previous chapter introduced the concept and key ideas for our self-adaptive hybrid multi-core. For self-adaptation, the focus was set to performance and temperature. The employed models and algorithms for performance management and thermal management will be presented in the next two chapters separately.

If you knew Time as well as I do, you wouldn't talk about wasting IT. It's HIM.

The Hatter, *Alice's Adventures In Wonderland*

CHAPTER 5

Performance Management

This chapter describes how self-adaptation can be applied on hybrid multi-cores for the purpose of performance management. The basis for this are the performance model and the self-adaption algorithms. For a realistic evaluation, the algorithms are applied on a real-world case study, a video object tracker, which is also presented in this chapter. The corresponding experimental results can be found in Chapter 7.

5.1 Performance Model

Formally, we denote the set of threads as T, a distinct thread as T_i, the set of cores as C, an individual core as C_j, and a specific thread instance being executed on a specific core as $T_{i,j}$. Thread instances generally show differing performance on different (hybrid/heterogeneous) cores, e.g., threads with data-parallel computations can be more efficiently mapped into reconfigurable logic. General speedup factors of an individual core, which holds for any kind of thread, cannot be given. While data-parallel computations can provide good speedups when mapped to reconfigurable logic, sequential algorithms usually show better performance on processors. Hence, the speedup factors are thread-dependent.

Let $s_{i,j}$ be the specific speedup of a (worker) core C_j executing the thread instance of T_i compared to the thread instance being executed on the master core. The speedup of the thread instance on the master core is always set to $s_{i,j=\text{master}} = 1$. We assume that the workload of thread T_i is distributed among

67

the cores executing instances of T_i proportionally to their speedup values $s_{i,j}$. Specifically, if the master core j executes a thread instance $T_{i,j}$ and a worker core j' executes another thread instance $T_{i,j'}$ with the speedup value $s_{i,j'} = 1.5$, we assume that the master core executes $\frac{2}{5}$ of the entire workload and the worker core $\frac{3}{5}$ of the workload, see Figure 5.1.

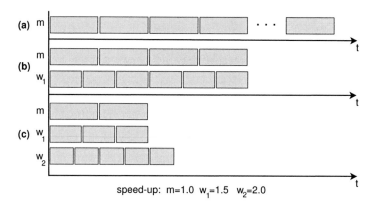

$$\text{speed-up: } m=1.0 \quad w_1=1.5 \quad w_2=2.0$$

Figure 5.1: The (a) mapping shows the scenario where ten work packages of a thread T_i are computed by the master m. In contrast, the (b) mapping shows the case where the worker core w_1 with $s_{i,1} = 1.5$ computes $\frac{3}{5}$ of the workload. For the (c) mapping, the worker core w_2 with $s_{i,2} = 2.0$ joins in and the workload is distributed among all three cores.

The main purpose of this assumption is to allow us to predict the application's execution time when adding or removing another thread instance. With $e_{i,master}$ as the *measured* fraction of T_i's execution time on the master processor under the current thread-to-core mapping, we can model a *hypothetical* execution time (or workload) $W(T_i)$ for a mapping, which would map T_i only on the master processor (and not on any worker core):

$$W(T_i) = e_{i,master} \times \sum_{\forall j:T_i \in \mu(C_j)} s_{i,j} = e_{i,master} \times S_i \qquad (5.1)$$

Here, $\mu(C_j)$ represents the set of thread instances being executed on core C_j, and S_i denotes the estimated cumulated speedup of thread T_i. Note that because starting or stopping a thread instance changes both $e_{i,master}$ and S_i, $W(T_i)$ remains constant regardless of the chosen thread-to-core mapping. An additional thread instance $T_{i,j'}$ on a free core C'_j will lead to savings in the master processor's actual execution time, i.e. reward, for that thread of

$$reward_{i,j'} = e_{i,master} \frac{s_{i,j'}}{S_i + s_{i,j'}} \qquad (5.2)$$

Similarly, terminating a thread instance $T_{i,j'}$ on a worker core C'_j will lead to a likewise increase in the master processor's actual execution time, i.e. penalty, for that thread of

$$penalty_{i,j'} = e_{i,master} \frac{s_{i,j'}}{S_i - s_{i,j'}} \qquad (5.3)$$

The resulting system performances are denoted as p_{reward} and $p_{penalty}$, respectively.

5.2 Self-adaptation Algorithms

In this section we present two self-adaptive add/remove strategies for parallel multithreaded applications on hybrid multi-core systems, i.e. a video object tracker. The two adaptive thread partitioning (ATP) algorithms, ATP$_{bound}$ and ATP$_{budget}$, aim to satisfy soft real-time constraints while minimizing the required processing resources. ATP$_{bound}$ always aims at keeping the system performance above a lower bound. ATP$_{budget}$ tries to stay within a performance budget and to attain a predefined average performance, thereby trading lower resource usage for slightly higher rates of missing the lower performance bound.

For both algorithms, we assume that an instance of each thread being executed on the master processor at all times and that each worker core is assigned at most one thread instance in total. Initially, the application with all its threads being entirely executed on the master processor. The master processor also measures the total application performance at user-defined time intervals.

5.2.1 Performance Constraint with Lower Bound

Algorithm 5.1 presents the pseudo code for our bound-based self-adaptation technique ATP$_{bound}$, which dynamically adds or removes thread instances in order to fulfill a lower performance bound while minimizing the number of active cores. The algorithm is executed in user-defined time intervals and re-partitions the system if the application's current performance p is below or above the user-defined performance bound $\mathcal{P}$. The repartition is realized by either calling add_core() or remove_core(). The remapping is achieved by reconfiguration, i.e. the dynamic partial reconfiguration of a hardware slot for hardware threads or the activation/deactivation of a worker processor for software threads. In case the performance is below the bound, the master creates an additional instance of the thread on the core that promises the largest increase in performance. If the performance exceeds the bound $\mathcal{P}$, the master terminates the instance of the thread that will free a core but keep the application performance above $\mathcal{P}$, if such a thread exists.

Algorithm 5.1 Pseudo code for ATP$_{bound}$ [8]

Require: set T of active threads and their instances $T_{i,j}$, set C of cores, thread execution times $e_{i,master}$ measured on the master processor, current application performance p.

1: **procedure** ATP$_{bound}$
2:    **if** p $< \mathcal{P}$ **then**   ▷ Guarantee lower bound
3:       add_core()
4:    **else if** p $> \mathcal{P}$ **then**   ▷ Check if a thread instance can be terminated
5:       remove_core()
6:    **end if**
7: **end procedure**

In Algorithm 5.2, the function add_core() computes for all threads $T_i \in T$ the possible rewards in execution time reward$_{i,j'}$ in case an additional thread instance is started on an idle core. The function only considers idle worker cores (line 4). Here, $\mu(C_j)$ defines the set of thread instances being executed on core C_j. Therefore, an idle core is characterized by $\mu(C_j) = \emptyset$. The reward is set to 0 for every other core in line 7. Then, in line 10, the function selects the thread instance $T_{i,j}$ that maximizes the estimated performance. Therefore, we estimate the performance $p_{reward_{i,j}}$ assuming that the execution time of the thread instances running on the master core can be reduced by $reward_{i,j}$. In case of several threads promising the same performance, we break ties by randomly choosing a thread. Finally, in lines 12-15, we create a new thread instance $T_{i',j'}$ and update the cumulated estimated speedup $S_{i'}$ and $\mu(C_{j'})$. If there is no idle worker core, the maximum achievable reward is 0. In this case, the current thread-to-core mapping remains unchanged.

The function remove_core() in Algorithm 5.3 operates in a similar way. It terminates the thread instance $T_{i,j}$ that promises the lowest increase in execution time, i.e. penalty. We do not deactivate a core, if the estimated performance drops below the user-defined bound $\mathcal{P}$ (line 35). In line 4 of Algorithm 5.3, we make sure that we only consider threads being executed on worker cores ($T_i \in \mu(C_j) \wedge C_j \neq master$).

Algorithm 5.2 Pseudo code for ATP$_{bound}$: add_core() [8]

1: **procedure** ADD_CORE
2:     **for** $i = 1, \ldots, |T|$ **do**   ▷ For all (thread,core) tuples
3:         **for** $j = 1, \ldots, |C|$ **do**
4:             **if** $\mu(C_j) = \emptyset \wedge C_j \neq master$ **then**
5:                 $reward_{i,j} = e_{i,master} \frac{s_{i,j}}{S_i + s_{i,j}}$ ▷ Compute individual rewards
6:             **else**
7:                 $reward_{i,j} = 0$
8:             **end if**
9:         **end for**
10:     **end for**
11:     $i', j' = \arg\max_{i,j}\{reward_{i,j}\}$ ▷ Select maximum reward
12:     **if** $reward_{i',j'} > 0$ **then**
13:         create $T_{i',j'}$ ▷ Create new thread instance
14:         $S_{i'} = S_{i'} + s_{i',j'}; \mu(C_{j'}) = \{T_{i'}\}$ ▷ Update cumulative speedup
15:     **end if**
16: **end procedure**

Algorithm 5.3 Pseudo code for ATP$_{bound}$: remove_core() [8]

1: **procedure** REMOVE_CORE
2:     **for** $i = 1, \ldots, |T|$ **do**   ▷ For all (thread,core) tuples
3:         **for** $j = 1, \ldots, |C|$ **do**
4:             **if** $T_i \in \mu(C_j) \wedge C_j \neq master$ **then**
5:                 $penalty_{i,j} = e_{i,master} \frac{s_{i,j}}{S_i - s_{i,j}}$ ▷ Compute individual penalties
6:             **else**
7:                 $penalty_{i,j} = \infty$
8:             **end if**
9:         **end for**
10:     **end for**
11:     $i', j' = \arg\min_{i,j}\{penalty_{i,j}\}$ ▷ Select minimum penalty
12:     **if** $penalty_{i',j'} < \infty \wedge p_{penalty_{i,j}} > \mathcal{P}$ **then**   ▷ Respect lower bound
13:         terminate $T_{i',j'}$ ▷ Terminate thread instance
14:         $S_{i'} = S_{i'} - s_{i',j'}; \mu(C_{j'}) = \emptyset$ ▷ Update cumulative speedup
15:     **end if**
16: **end procedure**

5.2.2 Performance Constraint with Lower and Upper Bound

Algorithm 5.4 presents the pseudo code for our budget-based add/remove self-adaptation technique ATP_{budget}. The algorithm is similar to the ATP_{bound} algorithm but differs in its objective. While the ATP_{bound} algorithm aims at keeping the application's performance above a performance bound, the ATP_{budget} algorithm seeks to stay inside a performance budget. This budget is spanned around a user-defined target performance $\mathcal{P}_{target}$, from which an user-defined aberration is allowed. Occurring deviations from the performance budget are weighted equally in the sense that mappings that are x times faster are equivalent to those that are x times slower.

Algorithm 5.4 Pseudo code for ATP_{budget} [8]

Require: set T of active threads and their instances $T_{i,j}$, set C of cores, thread execution times $e_{i,master}$ measured on the master processor, current application performance p.

1: **procedure** ATP_{budget}
2:     **if** p $< \mathcal{P}_{lower}$ **then**   ▷ Respect lower bound
3:         add_core()
4:     **else if** p $> \mathcal{P}_{upper}$ **then**   ▷ Respect upper bound
5:         remove_core()
6:     **end if**
7: **end procedure**

The algorithm is executed in user-defined time intervals and re-partitions the system only if the application's current performance p is outside the specified budget $[\mathcal{P}_{lower}, \mathcal{P}_{upper}]$ by either calling add_core() or remove_core(), lines 2–6. This reduces the number of function calls of add_core() or remove_core(). In case the performance drops below $\mathcal{P}_{lower}$, the master creates an additional thread instance on the core that promises either meeting the desired performance budget, if possible, or else the largest increase in performance. When the performance exceeds the upper threshold $\mathcal{P}_{upper}$, the master core terminates the instance of the thread that will lead to the reduction that is as close as possible to the desired performance target.

Lines 11-12 of Algorithm 5.5 feature a modified version of the add_core() function, which selects the thread-to-core mapping that approximates $\mathcal{P}_{target}$ best. Corresponding modifications can be found for the function remove_core() in lines 11–12 (Algorithm 5.6). The add_core() and remove_core() functions of ATP_{budget} and ATP_{bound} have an execution time of $O(|C||T|)$.

Algorithm 5.5 Pseudo code for ATP$_{budget}$: add_core() [8]

1: **procedure** ADD_CORE
2:     **for** $i = 1, \ldots, |T|$ **do**   ▷ For all (thread,core) tuples
3:         **for** $j = 1, \ldots, |C|$ **do**
4:             **if** $\mu(C_j) = \emptyset \wedge C_j \neq master$ **then**
5:                 $reward_{i,j} = e_{i,master} \frac{s_{i,j}}{S_i + s_{i,j}}$ ▷ Compute individual rewards
6:             **else**
7:                 $reward_{i,j} = 0$
8:             **end if**
9:         **end for**
10:     **end for**
11:     $i', j' = \arg\min_{i,j} \frac{\max(\mathcal{P}_{target}, \mathcal{P}reward_{i,j})}{\min(\mathcal{P}_{target}, \mathcal{P}reward_{i,j})}$
12:     **if** $\frac{\max(\mathcal{P}_{target}, \mathcal{P}reward_{i',j'})}{\min(\mathcal{P}_{target}, \mathcal{P}reward_{i',j'})} < \frac{\max(\mathcal{P}_{target}, p)}{\min(\mathcal{P}_{target}, p)}$ **then**   ▷ Select best reward
13:         create $T_{i',j'}$ ▷ Create new thread instance
14:         $S_{i'} = S_{i'} + s_{i',j'}; \mu(C_{j'}) = \{T_{i'}\}$ ▷ Update cumulative speedup
15:     **end if**
16: **end procedure**

Algorithm 5.6 Pseudo code for ATP$_{budget}$: remove_core() [8]

1: **procedure** REMOVE_CORE
2:     **for** $i = 1, \ldots, |T|$ **do**   ▷ For all (thread,core) tuples
3:         **for** $j = 1, \ldots, |C|$ **do**
4:             **if** $T_i \in \mu(C_j) \wedge C_j \neq master$ **then**
5:                 $penalty_{i,j} = e_{i,master} \frac{s_{i,j}}{S_i - s_{i,j}}$ ▷ Compute individual penalties
6:             **else**
7:                 $penalty_{i,j} = \infty$
8:             **end if**
9:         **end for**
10:     **end for**
11:     $i', j' = \arg\min_{i,j} \frac{\max(\mathcal{P}_{target}, \mathcal{P}penalty_{i,j})}{\min(\mathcal{P}_{target}, \mathcal{P}penalty_{i,j})}$
12:     **if** $\frac{\max(\mathcal{P}_{target}, \mathcal{P}penalty_{i',j'})}{\min(\mathcal{P}_{target}, \mathcal{P}penalty_{i',j'})} < \frac{\max(\mathcal{P}_{target}, p)}{\min(\mathcal{P}_{target}, p)}$ **then**   ▷ Select best penalty
13:         terminate $T_{i',j'}$ ▷ Terminate thread instance
14:         $S_{i'} = S_{i'} - s_{i',j'}; \mu(C_{j'}) = \emptyset$ ▷ Update cumulative speedup
15:     **end if**
16: **end procedure**

5.3 Streaming Case Study

Self-adaptive multi-cores allow for the transparent distribution of execution units between hardware cores and worker processors by simply instantiating the corresponding threads, as described in Chapter 4. This makes hybrid multi-cores an ideal target for applications composed of several classes of remote threads for which the actual distribution and number of threads in each execution context (hardware or software) depends on input data characteristics and other run-time parameters. For performance management, we have defined streaming applications as target applications on the hybrid multi-cores, see Section 3.4.1.

As an example for streaming case studies, we studied sequential Monte Carlo (SMC) methods. SMC methods are applied when the system state is unknown and has to be estimated based on probabilistic distributions. The probabilistic distribution is approximated by a set of estimates that can be processed in parallel because they are independent of each other. Furthermore, SMC methods usually contain different stages, which can be pipelined.

An example class of applications that profits from such a dynamic execution environment is the on-line state estimation through stochastic methods. We have developed a framework for Sequential Monte-Carlo Methods [3–5, 7, 8], also called particle filters, which are applicable in a wide range of applications, such as network packet processing, mobile communications, navigation systems, or visual object tracking.

Section 5.3.1 first provides background information on SMC methods, before Section 5.3.2 introduces the multithreaded particle filter framework for hybrid multi-cores. Finally, Section 5.3.3 describes the video object tracker as a real world case study that implements the particle filter framework.

5.3.1 Sequential Monte Carlo (SMC) Methods

Sequential Monte Carlo methods are a stochastic approach to track the system state x_t over time in a dynamic environment, where t denotes a discrete point in time. It is assumed that the system state cannot be observed directly but that the system continuously receives noisy measurements y_t at discrete time steps t. Since the system state x_t is unknown, a random variable X_t describes its probability distribution. To reduce the computational overhead, the probability distribution of X_t is approximated using a (fixed) number of discrete samples x_t^i, which are often referred to as particles. SMC methods rely on two models, which have to be adjusted to the application field of the system: the `system model` and the `measurement model`. The system model estimates the probability distribution $p(X_t|X_{t-1})$ for the system state X_t for a given previous system state X_{t-1}. The system model describes a Markov process of first order. Usually, the initial system state $x_{t=0}$ and therefore the initial probability distribution $p(X_t = x_0)$ is known. In a next step, the measurement model $p(Y_t = y_t|X_t)$ computes the likelihood for each particle x_t^i that the measurement y_t was observed in the system state x_t^i.

Based on the system and measurement models, we are interested in **predicting** the current state based on past measurements, $p(X_t|y_1, \cdots, y_{t-1})$, or in **updating** the current state prediction based on the latest measurement, $p(X_t|y_1, \cdots, y_t)$. Sequential Monte Carlo methods approximate these distributions by drawing a large number of samples from them. However, as these distributions are typically unknown, we cannot directly draw samples from them. Several techniques have been proposed to circumvent this problem, among them the **importance sampling**. When the estimates for the system state have to be computed on-line while the state changes, as it is the case for tracking applications, the **sequential importance sampling** (SIS) and **sequential importance resampling** (SIR) methods can be applied, which presents a recursive approach to approximate the desired distributions. Our framework closely follows the SIR algorithm (Algorithm 5.7), as it is one of the most widely used sequential Monte Carlo methods.

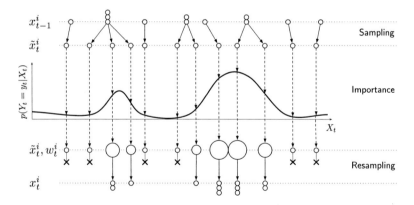

Figure 5.2: Sampling-importance-resampling algorithm: The figure illustrates one iteration of the algorithm. [3]

A single iteration of the SIR algorithm is depicted in Figure 5.2. Here, the state space is one-dimensional and particles are represented by circles. In a first step, the sampling stage, the new set of particles $\tilde{x}_t^i$ for the next time step t are sampled using the system model $p(X_t|X_{t-1} = x_{t-1}^i)$. The new set of particles $\tilde{x}_t^i$ defines the current probability distribution X_t. In a next step, the importance stage, all particles are weighted according to the measurement model $p(Y_t = y_t|X_t)$, which is depicted as a two-dimensional function. In Figure 5.2, the circle radii correspond to the particles' weight. Particles $\tilde{x}_t^i$ with a high likelihood that current measurement y_t has been received in their state are assigned high weights w_t^i. In a final step, the resampling stage, particles with high weights are duplicated and particles with low weights are eliminated. Note that the number of particles stays constant at all stages of the SIR algorithm. After resampling, the unweighted particles x_t^i approximate the weighted particles $\tilde{x}_t^i, w_t^i$.

As can be derived from Figure 5.2, the SIR algorithm replicates all particles that form a likely prediction of the state while eliminating particles that do not match the current measurement. This process can be viewed as a filter that tracks several state hypotheses at once, filtering out hypotheses that are not supported by the measurements. After each time step, the most likely match can be extracted from the probability distribution approximated by the particles, i.e., by selecting the particle with maximum weight or computing the average over all particles.

Algorithm 5.7 Sampling-Importance-Resampling algorithm. (source: [23])

1: $[\{x_t^i, w_t^i\}_{i=1}^N] = $ SIR $[\{x_{t-1}^i, w_{t-1}^i\}_{i=1}^N, y_t]$
2: **for** $i = 1, \dots, N$ **do**
3:      Draw $x_t^i \sim p(X_t|X_{t-1} = x_{t-1}^i)$ $\triangleright$ Sampling
4:      $w_t^i = p(Y_t = y_t|X_t = x_t^i)$ $\triangleright$ Importance
5: **end for**
6: $W = \sum_{i=1}^N w_t^i$ $\triangleright$ Calculate total weight
7: **for** $i = 1, \dots, N$ **do**
8:      $w_t^i = w_t^i/W$ $\triangleright$ Normalize weights
9: **end for**
10: $[\{x_t^i, w_t^i, -\}_{i=1}^N] = $ RESAMPLE $[\{x_t^i, w_t^i\}_{i=1}^N]$ $\triangleright$ Resampling

---

Algorithm 5.7 presents the pseudo code of the SIR algorithm for N particles. In lines 2–5, the particles x_t^i are first drawn according to the prediction $p(X_t|X_{t-1} = x_{t-1}^i)$ and then weighted according to the likelihood $p(Y_t = y_t|X_t = x_t^i)$. As the particles are independent from each other, the for-loop can be parallelized. Furthermore, the sampling and importance stages can be pipelined. For the resampling stage, see line 10, the particle weights have to be normalized, see lines 6–9. In most cases, the current state prediction is either the particle that has the highest weight or the (weighted) average of all particles.

Algorithm 5.8 Resampling algorithm. (source: [23])

1: $[\{x_t^i, w_t^i, -\}_{i=1}^N] = \text{RESAMPLE } [\{x_t^i, w_t^i\}_{i=1}^N]$ ▷ Resampling
2: $c_0 = 0$ ▷ initialize CDF
3: **for** $i = 1, \ldots, N$ **do**
4:     $c_i = c_{i-1} + w_t^i$ ▷ Construct CDF
5: **end for**
6: $i = 1$ ▷ Start at the bottom of CDF
7: $u_i \sim U[0, \frac{1}{N}]$ ▷ Draw a starting point
8: **for** $j = 1, \ldots, N$ **do**
9:     $u_j = u_1 + \frac{1}{N}(j - 1)$ ▷ Move along the CDF
10:     **while** $u_j > c_i$ **do**
11:        $i = i + 1$
12:     **end while**
13:     $x_t^{j^*} = x_t^i$ ▷ Assign sample
14:     $w_t^j = \frac{1}{N}$ ▷ Assign weight
15:     $i^j = i$ ▷ Assign parent
16: **end for**

Algorithm 5.8 presents the resampling algorithm, which uses a cumulative probability distribution function (CDF) of p. Let c_i be the subtotal of the particle weights up to the i-th particle, see lines 2–5. The algorithm randomly selects a starting point between 0 and $\frac{1}{N}$ of the resampling function u_1, see lines 6–7. The for-loop in lines 8–16 calculates for every particle x_t^j the number of its replications. For this, the current resampling function value u_j is set to $u_1 + \frac{1}{N}(j - 1)$, such that all resampling function values are equidistant. Then, the j-th particle is replicated until its CDF value c_j exceeds the current resampling function value u_i.

Bolić et al. [30] introduced residual systematic resampling (RSR) for particle filters. RSR calculates the replication factors r_j for every particle instead of entire particle sets. Thus, only one generation of particles needs to be stored into memory. Note that the resampling stage and the sampling stage need to be updated to apply RSR. If u_1 and all particle weights w_t^i are known beforehand, the resampling stage can be parallelized as well, as shown in [24]. We follow both approaches to provide multiple threads for the resampling stage. A more thorough discussion of the theoretical foundations of SMC methods can be found in [44].

In this thesis, we implemented a particle filter as a case study on a self-adaptive hybrid multi-core. Therefore, we developed a generic multithreaded framework, which utilizes reconfigurable hardware, i.e. hardware cores, to accelerate the filter's performance. Next to our particle filter framework, accelerating particle filter algorithms with reconfigurable hardware has also been addressed by several other research groups in previous years. For instance, Athalye et al. [24] developed methods and architectures for accelerating the resampling step of the SIR algorithm while at the same time reducing the memory requirements for hardware implementations. Our framework adapts their technique for parallelizing the resampling phase.

Furthermore, Sankaranarayanan et al. [121] presented a flexible hardware architecture for SMC methods that uses density sampling techniques from the more general domain of Monte Carlo Markov chain algorithms. This allows them to drop the resampling step, which poses scalability and efficiency issues when implemented in hardware. However, the approach does not show significant improvements in quality over traditional SIR filters. As we resolve some of the efficiency issues by implementing the resampling step in software, we chose not to adapt this method.

Finally, Saha et al. [119] presented a parameterizable framework for the hardware implementation of particle filters, which bears some similarity to our approach in that it provides an interface for the model definition of a particle filter. However, their proposed framework targets a static hardware-only implementation of the filter and thus significantly differs from our flexible multithreaded HW/SW approach. Furthermore, their static approach does not support any on-line self-adaptation techniques.

5.3.2 Multithreaded SMC Framework

All particle filters using the SIR algorithm rely on the same underlying algorithmic structure. Hence, a substantial part of the functionality, code, and – in the case of hybrid CPU/FPGA systems – hardware circuitry can be re-used supported by a framework-based design approach. Our particle filter framework takes care of common tasks shared by all SIR implementations, such as data transfer and control flow and lets the designer focus on the application-specific tasks, such as system and measurement modeling.

The characteristic feature of our particle filter framework is the use of multithreaded programming across the hardware/software boundary. The combination of control-centric and data-oriented processing inherent in particle filters closely matches the target platforms and capabilities of our multithreaded operating system ReconOS, which is used for its implementation.

Figure 5.3 depicts the structure of our multithreaded particle filter framework, which is based on a SIR filter. Compared to a regular SIR filter, the framework

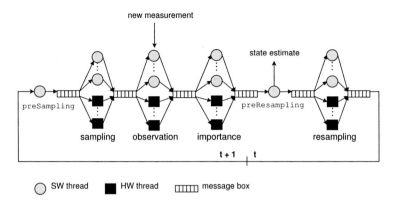

new measurement

state estimate

preSampling

preResampling

sampling   observation   importance   resampling

t + 1   t

○ SW thread   ■ HW thread   ⊓⊓⊓⊓ message box

Figure 5.3: Structure of a SIR filter implementation [4]

provides an additional filter stage, the observation stage, which is positioned in-between the sampling and the importance stage. The observation stage provides a particle-specific preprocessing of the measurement data, which we found could be useful for many applications that require a computational intensive preprocessing. All four stages can be implemented using software threads, hardware threads, or both. As the particles can be processed independently, multiple threads can be instantiated in order to parallelize the particle processing at stage level. Furthermore, the first three stages can be pipelined, which permits a parallel execution of different stages. The resampling stage requires information of all particles, hence the fourth stage cannot be pipelined with the previous stages.

The particle filter was implemented using the ReconOS programming model. The particle set is partitioned into chunks of user-defined size. The filter stages employ message boxes for inter-thread communication where a message contains the identifier of a particle chunk. Partitioning the particles into chunks is a compromise between optimally balancing the workload for the thread instances of a stage, on the one side, and minimizing the communication overhead between stages, on the other side.

As shown in Figure 5.3, a software thread named `preSampling` precedes the sampling stage. This thread retrieves new measurements by using the application-specific *receive_new_measurements* function and prepares the sampling stage by reassigning particles to chunks. This is necessary, because the resampling stage of the previous iteration replicates some particles and deletes others, leaving the chunks non-uniformly populated with particles. Before particles can be weighted at the importance stage, an observation has to be extracted from the measurement for each particle in the observation stage. Finally, before resampling, the preResampling thread synchronizes the data flow of all particles to normalize their weights. Additionally, a user function *iteration_done* is executed to take care of application-specific tasks once per iteration, such as transferring of estimation results to a display, memory, or input/output, as well as hardware/software repartitioning of the threads in the filter stages. The latter is useful when the stage thread's performance is data-dependent and thus changes during runtime.

The application-specific system and measurement models are captured by the functions *prediction*, *extract_observation* and *likelihood*. For a software implementation, the user simply fills out C function templates provided by the framework, which are linked into the corresponding software threads of the sampling, observation and importance stages. When creating hardware versions of these functions, the user writes specialized entities in VHDL, which are then embedded into the hardware threads provided by the framework. Figure 5.4 illustrates the composition and interactions between functions and modules, which are provided by the user and the framework to assemble a complete application. The set of application-specific functions is listed in Table 5.1.

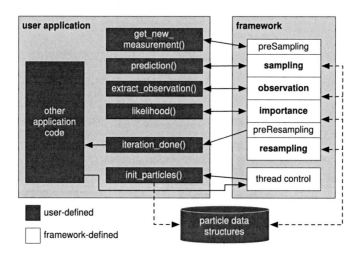

Figure 5.4: System composition: application and framework functions [4]

For the individual filter stages, an arbitrary number of software and hardware threads for each filter iteration can be chosen at system start. This number can

Table 5.1: User functions for customizing the particle filter framework

function	description
init_particles	initializes the particles
prediction	moves a particle using the system model
get_new_measurement	retrieves a new measurement each filter iteration
extract_observation	extracts an observation for a particle from the current measurement
likelihood	weights a particle using the measurement model
iteration_done	called once per iteration; selects best particle updates HW/SW partitioning

be updated dynamically at run-time with the only constraint that each stage must be represented by at least one thread at all times. Usually, the number of thread instances per stage is limited by the available system resources, such as the number of hardware and software cores. Next to the four filter stages, there are further threads that are all implemented in software. These threads are mostly control-dominated and are therefore a poor candidate for a hardware implementation.

Table 5.2: Framework functions for initialization and execution control

function	description
create_particle_filter	creates and initializes the particle filter structure
init_reference_data	initializes reference data for *likelihood* function
set_sample_hw/sw	sets number of HW/SW threads for sampling
set_observe_hw/sw	sets number of HW/SW threads for observation
set_importance_hw/sw	sets number of HW/SW threads for importance
set_resample_hw/sw	sets number of HW/SW threads for resampling
start_particle_filter	starts the particle filter

For instance, one software thread initializes the data structures and the hardware/software partitioning of the particle filter. To update the hardware/software partitioning of the particle filter, the framework provides software functions that set the numbers of hardware and software threads for each stage. The hardware/software partitioning can be updated once per filter iteration. Table 5.2

shows an overview of the framework functions that can be called by the designer to create and initialize the particle filter and to update the hardware/software partitioning.

Altogether, the framework provides all necessary data structures, a well-defined control flow, and a flexible execution environment, which allows a dynamic adaptation of the hardware/software partitioning for particles filters over time.

5.3.3 Video Object Tracking

Using this framework, we implemented a video object tracker as a real-world case study, which tracks a moving object inside a video sequence. The user selects the initial position of an object and its approximated outline by defining a bounding box in the first video frame. The color histogram of the pixels inside the bounding box determines the object's reference histogram, which is used by the video object tracker to identify the object in the video sequence. It is the task of the particle filter to track the position, speed, and the scale of the tracked object. A tracking example for a soccer sequence is shown in Figure 5.5.

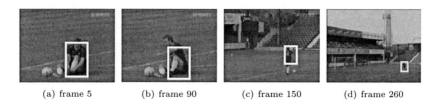

| (a) frame 5 | (b) frame 90 | (c) frame 150 | (d) frame 260 |

Figure 5.5: Object tracking in a video sequence (soccer) [3]

During the sampling stage, the filter moves the particles according to their current velocity and adds some noise to achieve a better distribution as well as to model uncertainty. Then, the observation stage computes color histograms for all particles using the current video frame, which serves as the measurement. In the importance stage, the particles are weighted according to a their histograms, which are compared against the reference histogram of the object in the likelihood function.

The likelihood function is an exponential function where the exponent describes how closely the particle's histogram matches the object's reference histogram. The particle with the highest weight is then drawn as the most likely new system state, shown as a white outline in Figure 5.5. More details about the object tracking application can be found in [4].

The video object tracking application is a prime example to demonstrate self-adaptive thread-to-core mapping for two reasons. First, the required processing power strongly depends on the actual contents of the video frames and can vary in a wide range. Second, the data-parallel application lends itself to a thread structure where executing more instances of the same thread helps to increase the performance, see Figure 5.3.

5.4 Chapter Conclusion

In this chapter, we defined the performance model for self-adaptive hybrid multi-cores. This model assumes that each application thread can have multiple instances, which can be dynamically created and terminated at run-time. In order to limit the reconfiguration overhead, the self-adaptability of the multi-core is limited by the restriction that for each adaptation only one thread instance can be created or terminated in total. The self-adaptation algorithms, ATP_{bound} and ATP_{budget}, were presented. The first algorithm, ATP_{bound}, adapts the thread-to-core mapping to meet a lower user-defined performance bound using as few cores as possible, i.e. to save power. In contrast, the second algorithm, ATP_{budget}, intends to keep the performance inside a user-defined performance interval, again with the goal to utilize as few cores as possible.

Furthermore, this chapter presented a multithreaded particle filter framework that was designed to simplify the development of particle filter-based streaming applications on self-adaptive hybrid multi-cores. A video object tracker was presented as a real-world case study whose performance, in this case tracking throughput, strongly depends on the input data. Thus, the video object tracker provides an excellent experimental environment to evaluate the effectiveness of the self-adaptation techniques for performance management on a real-world system.

The next chapter will introduce the thermal model and self-adaptation techniques for thermal management on self-adaptive hybrid multi-cores. Furthermore, it will present a novel self-calibration technique for ring oscillator-based temperature sensors and dedicated heat-generation cores for creating spatial temperature differences. Finally, a temperature simulator will be discussed, which will be used to evaluate the proposed self-adaptation techniques.

'Are five nights warmer than one night, then?' Alice ventured to ask. *'Five times as warm, of course.'*
'But they should be five times as COLD, by the same rule—'
'Just so!' cried the Red Queen. *'Five times as warm, AND five times as cold—just as I'm five times as rich as you are, AND five times as clever!'*

Lewis Carroll, *Through The Looking Glass*

CHAPTER 6

Thermal Management

This chapter addresses autonomous thermal management techniques for hybrid self-adaptive multi-cores. The chapter presents a study of ring oscillator-based temperature sensors, which are used for fine-grained temperature measurements. A novel self-calibration method is presented for the sensors.

The thermal management techniques, which are developed as part of this thesis, target future FPGA technologies that have to cope with a dramatic increase in temperature. In this thesis we investigate if high temperatures and (spatial) thermal imbalances can already be generated and observed on today's FPGAs. Therefore, this chapter presents dedicated heat-generating cores to purposefully generate high temperatures in certain regions of the chip.

In order to predict the thermal heat flow of the FPGA the system learns its thermal model at run-time. This chapter defines a thermal model of the FPGA chip and the applied learning strategy. Using this thermal model, the thermal effects of thread migrations on the hybrid multi-core can be estimated or predicted in order to select the best possible thread migration to balance the chip temperature. The prediction model and the self-adaptation algorithms for autonomous thermal management are presented in this chapter. Finally, a temperature simulator is introduced, which is used to evaluate the effectiveness of the temperature-driven self-adaptation algorithms.

6.1 Measuring Temperature on Today's FPGAs

In contrast to performance measurements, fine-grained temperature measurements are not yet supported by today's FPGA platforms. Indeed the vendors recognize the raising need for temperature measurements due to continuously shrinking microelectronic device structures and therefore integrate one or two temperature diodes on the chip. However, the number of temperature sensors is too low for allowing fine-grained thermal management on an FPGA-based hybrid multi-core. In this thesis, we assume that we have a regular sensor grid on the FPGA. To realize this sensor grid on today's FPGAs, we studied ring oscillators as additional temperature sensors. A ring oscillator based temperature sensor can be mapped to every slice of an FPGA, which are well-distributed across the die. Hence, we are able to build the sensor grid as defined in Section 3.4.2 using ring-oscillator based temperature sensors. The sensor architecture was already presented in Section 4.6. However, in this section an evaluation methodology is presented to experimentally define the optimal measurement period and the best sensor layout in order to obtain a good trade-off between maximizing the temperature resolution and minimizing the measurement noise.

6.1.1 Temperature Sensor: Design Space Exploration

In the last decade, many researchers employed ring oscillators as temperature sensors on FPGAs [108, 148, 170]. The research on ring oscillator-based temperature sensors includes sensor placement [104], sensor calibration [6, 89] and workload effects on the measurement accuracy [123]. While numerous publications have presented ring oscillator designs for temperature measurements, a detailed study of the ring oscillator design space is still missing.

Hence, this thesis introduces metrics for comparing the measurement quality and area efficiency of ring oscillators and a methodology for the experimental evaluation of a broad range of sensor designs. The measurement quality of a sensor is defined as its resolution divided by its noise. Using this definition, a good sensor has a high temperature resolution while only suffering from minor noise at the same time. We target modern Xilinx FPGAs that contain a built-in thermal diode, which we use for calibration and evaluation purposes. The examined designs differ in the ring oscillator size, slice utilization and in routing. We evaluate noise and measurement accuracy of the sensors and investigate the influence of a number of design parameters.

In the design of the temperature sensor there are a number of parameters that need to be defined for implementation. The *measurement period* t_m is the time interval in which we sample the ring oscillator's output signal. While a longer

t_m should lead to more accurate results, issues like self-heating, area allocated to the oscillation counter and the ability to obtain a time series of sensor readings with high temporal resolution, constrain the measurement period. The *number of inverters* and other delay elements such as *latches* have a significant effect on the measurement quality of the ring oscillator. And finally, the placement of the ring oscillator elements, as well as routing have an impact on the sensor's measurement quality.

To evaluate our sensors we need an independent way to determine the on-chip temperature. For this, we use the built-in system monitor of the FPGA that employs a thermal diode with an accuracy of $\pm 4°C$ as specified by the manufacturer [157].

For our evaluation of the different sensor designs, we establish an evaluation function that takes into account the sensor resolution σ_v and the sensor noise, expressed as the standard deviation σ_c of the sensor readings. We calculate the resolution of the sensor as

$$\sigma_v := \frac{S_{max} - S_{min}}{\tau_{max} - \tau_{min}} \tag{6.1}$$

where S_{max} and S_{min} are the number of oscillations measured in the time interval t_m at maximum and minimum temperatures τ_{max} and τ_{min}, respectively. This value indicates how much the sensor count changes at a given temperature difference.

Sensor noise is determined over a time series of n single measurements at constant temperature with sensor readouts S_i and an average of $\overline{S}$.

$$\sigma_c := \max \left\{ \sqrt{\frac{1}{n} \sum_{i=1}^{n} (S_i - \overline{S})^2}, 0.5 \right\} \tag{6.2}$$

Since we expect a quantization noise of $\sigma = 0.5$ the value σ_c is clamped at 0.5 for low-noise sensors that happen to encounter little quantization error. We can then calculate the measurement quality G of the sensor, which can be seen as the sensor's signal to noise ratio and gives us a good indication of how well the sensor balances noise versus resolution.

$$G := \frac{\sigma_v}{\sigma_c} \tag{6.3}$$

6.1.2 Self-calibrating Technique

In order to get a good sensor coverage of the FPGA, we partition the FPGA into a regular grid of tiles and place a sensor at the center of each tile. Thus, the sensors form a sensor grid where all sensors are connected to the same monitoring unit. The monitoring unit is connected to the main CPU using the PLB bus. The main CPU can access every sensor of the monitoring core. An example sensor grid can be seen in Figure 6.1.

When a measurement is triggered by software each sensor performs a measurement as described in Section 4.6. The individual counter values can be accessed by the main processor using the PLB bus. These counter values S can then be translated to temperatures using calibrated translation functions $\eta(S)$. These translation functions $\eta(S)$ are set using a novel self-calibration technique that will be discussed in the following.

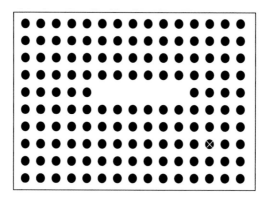

Figure 6.1: Example sensor layout: 15×10 sensor grid mapped on a Virtex-6 LX240T FPGA. The white box in the center is not reconfigurable. [6]

We aim at balancing the on-chip temperature distribution of the FPGA-based hybrid multi-core not only in simulations but also in experiments. For this purpose there is a need for cores that can generate significant spatial thermal differences already on today's FPGAs. Thus, we have developed regional heat-generating circuits (heaters), which we can map to certain FPGA regions using area constraints. These heaters can also be used to calibrate the temperature sensors because they enable the system to manipulate the chip temperature and, thus, can self-calibrate the sensors. Figure 6.2 shows an example positioning of such heaters on an FPGA.

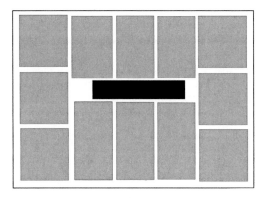

Figure 6.2: Example heater core layout: 12 regional heaters mapped on a Virtex-6 LX240T FPGA. The black box in the center is not reconfigurable. [6]

All heaters are connected to a system bus and can be activated independently from a processor. For calibration, we look for a uniform activity inside the FPGA. In our experiments, we have toggled the flip flops inside the heater's region at a certain frequency. For instance, using an Xilinx Virtex-6 FPGA, we were able to generate spatial thermal differences of up to 6.5°C while activating a subset of the heaters, where each heater toggles 10,000 flip flops at 100 MHz.

To translate the measured frequencies of a ring oscillator-based sensor to temperatures, the system needs to calibrate the sensor. To apply our approach, the FPGA needs to contain a built-in thermal diode, which the system uses to calibrate the sensors. Such thermal diodes can be found in Xilinx Virtex-5 and Virtex-6 FPGAs.

According to [89, 148], the correlation between measured frequencies and temperatures is almost linear. Thus, we propose that the system calibrates a sensor by mapping distinct sensor frequency measurements to the corresponding temperature measurements of the thermal diode. For a sensor calibration, the on-chip temperature distribution should be balanced. Hence, the system only measures twice, (1) when the chip temperature τ_1 has not been affected by the heaters with the sensor count S_1 and (2) after the chip temperature τ_2 was increased by all heaters with the sensor count S_2.

Here, we assume that by activating all regional heaters, a uniform temperature increase can be generated on the chip. The system activates all heaters to heat up the chip and waits until the temperature reading of the built-in thermal diode stabilizes. Then, all heaters are deactivated again, before the second measurement is performed. Due to process variation different parts of the chip can suffer from dissimilar leakage currents and, thus, dissipate diverse amounts of power. However, using the proposed approach, the diverse power dissipations of different chip regions have no effect on the sensor calibration, because the heaters are deactivated before the measurement takes place. This is true, because the dynamic power of the heaters directly diminishes to zero after deactivation, whereas the chip temperature cools down slowly such that a significant temperature increase can be measured.

The system generates a linear frequency-temperature transformation function $\eta(S)$ that is defined by the above mentioned measurements. Each sensor has to be calibrated independently due to process variation. Equation 6.4 shows how the sensors can translate their sensor count values S to temperatures τ after they have performed the calibration phase to define the measurements (S_1,τ_1) and (S_2,τ_2). Once the individual translation functions are defined for each sensor, the system can measure temperature distributions on the chip.

$$\eta(S) := \frac{\tau_2 - \tau_1}{S_2 - S_1}(S - S_1) + \tau_1 \qquad (6.4)$$

6.2 Generating High Temperatures

When investigating thermal management techniques on real world systems, the need arises to purposefully generate heat in specific regions of the FPGA [6, 166] using dedicated circuits. Little consideration has been given to the design of such circuits in the past. Often, ad-hoc solutions involving flip flop pipelines are employed without further examination of possible alternatives. However, there are many different resources on modern FPGAs that may be integrated into a heat generator.

Hence, we performed a detailed examination of various heat-generating cores, examining a variety of different approaches using different on-chip resources and heat generation techniques, such as ring oscillators, LUT-FF-pipelines, shift registers and DSP blocks. Therefore, we investigate which resources of an FPGA are best-suited to generate a high temporal thermal difference.

We developed dedicated cores that have the single purpose of generating heat. In our experiments we studied different resources (LUTs, FFs, BRAMs, DSPs) separately and, in the case of LUTs and FFs, in combination. In order to create maximum heat, we intend to provoke as much toggling of our signals and/or our storage elements as possible. For most of our designs we connected our resource elements in pipelines in order to minimize the interconnect inside the core and thus enable higher clock frequencies. Since most of our cores are clock-controlled, we did not only evaluate the individual influence of each resource element but also the impact of the clock frequency on heat generation.

In our experiments, we could observe that on a Xilinx Virtex-5 LX110T FPGA, these heaters could increase the FPGA temperature in 700 seconds from $+3°C$ to $+134°C$. Altogether, eight different heaters were tested, most of them using varying frequencies between 100 MHz and 600 MHz. The best results could be observed for LUT-based ring oscillators whose toggling signals get distributed across the heater region utilizing as many routing resources as possible. The individual heaters are discussed in the following.

LUT pipeline

Figure 6.3 depicts a LUT-based pipeline. In all experiments 6-input LUTs (LUT6s) with six inputs (I) and one output (O) were used. One of the six input signals (I1) enables the LUT. The other five input signals are connected to the preceding LUT in the pipeline. The LUT is configured in such a way that it inverts one of the five input signals whenever the heater is enabled. The first LUT in the pipeline receives a toggling input signal, i.e. from a system clock. Here, the maximum clock frequency depends on the length of the LUT pipeline.

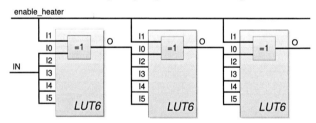

Figure 6.3: LUT-based pipeline [10]

LUT oscillator

LUTs can also be used as ring oscillators, where an odd number of inverters is connected to each other to form a ring. When the number of inverters is odd, the signal is unstable and toggles between '0' and '1'. The number of inverters and the delay of the interconnect define the toggling frequency. Since we intend to maximize the toggling frequency in order to create heat, we have used ring-oscillators with a single LUT, as depicted in Figure 6.4.

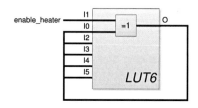

Figure 6.4: LUT-based ring oscillator [10]

A heater combines a number of these ring oscillators where all oscillating signals can be accessed by the central unit, and, thus, uses a lot of routing resources. This heater has shown the highest potential in heat-generating. It has been observed in experiments that the ring oscillators on their own do only generate a fraction of the high temperature, if there is no additional routing. The additional routing resources increase the load capacity of the circuit, which effects the dynamic power consumption of the circuit. As a result, most heat can be generated by toggling signals at maximum frequency using as much routing resources as possible (to maximize the load capacity).

SRL pipeline

Another possibility of using LUTs for the heat core is to use them as shift registers (SRLs). By cascading the SRLs it is possible to create one huge shift register, which is permanently shifting bits.

FF pipeline

One way to use FFs exclusively is to cascade them and build a shift register similar to the SRLs mentioned above. They have the same inputs, outputs and are clock-controlled with a clock enable signal. Beyond that, they have the same purpose, more precisely both pipelines are shift registers except that the FF can only store one bit compared to the 16 bit SRL in Virtex-5 FPGAs.

LUT-FF pipeline

Another way to exploit the given hardware components is to use the whole slice, instead of solely LUTs or FFs. In order to combine LUTs with FFs, we have designed a pipeline similar to the LUT pipeline (Figure 6.3) and the FF pipeline.

The difference is that between each pair of LUTs there is a FF interconnected, as can be seen Figure 6.5.

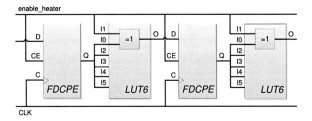

Figure 6.5: LUT and FF-based pipeline [10]

SRL-FF pipeline

An additional way to use LUTs and FFs is to make use of SRLs, instead of LUTs, and FFs. Therefore, each SRL's output signal is interconnected with the input signal of a FF. This creates a maximum capacity shift register for a given area.

BRAM pipeline

The BRAM-based heater is composed in a pipeline as well, using the Xilinx primitive FIFO36. Therefore the data input bus and the data output bus (with a width of 32 bit) are interconnected to each other. Once the first BRAM is filled with random data, it passes the first 32-bit word in it to the next BRAM. Hence, all BRAMs are permanently changing their memory contents.

DSP pipeline

Finally, multiple DSP blocks are cascaded and arranged in a pipeline. The output signal is passed through all DSP blocks by interconnecting it with the successor's input signal C. All input signals A and B of the DSP blocks are connected to global signals, which change their values with each clock cycle. Figure 6.6 depicts two strongly simplified segments of the DSP pipeline.

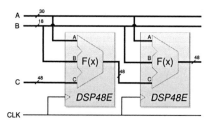

Figure 6.6: DSP-based pipeline [10]

6.3 Thermal Model

Using the self-calibrating regular sensor grid and the regional heaters, we are able to create and measure spatial thermal differences already on today's FPGAs. As we can create heat at specific regions on the FPGA and measure how this heat is transferred to the rest of the FPGA, we can learn the thermal model of an actual FPGA. This knowledge defines a foundation for later thermal management techniques. In contrast to performance management, the cores of the FPGA affect each other in thermal management. The temperature of the cores are not only affected by the threads being executed on the cores themselves, but also by the neighboring cores. For instance, the temperature of a core that idles can be increased when the surrounding / neighboring cores increase their temperature due to rising workload.

Hence, the system requires a thermal model of the specific FPGA that defines how the temperature distributes along the tiles of its regular grid, see Section 3.4.2. When the hybrid multi-core learns its thermal model and the thermal footprints of the individual (thread,core) tuples at run-time, the system can autonomously adapt its mapping in a smart manner in order to avoid unnecessary costly thread migrations that would be introduced by a simple trial-and-error method. Due to the hybrid and heterogeneous nature of the multi-core system, the thermal dynamics of thread migrations can hardly be overlooked. For this purpose, we introduce a thermal model of the underlying die.

The purpose of our thermal model lies not so much in accurately modeling physical reality given a set of parameters, as it is done for instance in design-time thermal analysis [73]. Instead we focus on an efficiently computable model with a small number of free parameters that can be easily found by an on-line learning algorithm.

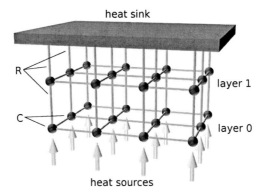

heat sink

R

layer 1

C

layer 0

heat sources

Figure 6.7: Thermal model: The nodes are arranged in two layers where each layer is a regular grid of nodes. Layer 0 receives heat input while layer 1 is connected to the heat sink. [6]

In order to model heat flow and temperature distributions we make use of the well known duality between thermal models and electrical RC-networks. Our model consists of two vertically arranged layers, L_0 and L_1. Each layer is a regular grid of nodes of width w and height h, which matches the layout of the temperature sensor grid. Each node i is identified by a coordinate $p(i) = (x,y)$, and the number $l \in \{0,1\}$ of its layer. A capacity $C(i)$ is assigned to each node. Nodes within a layer l are connected by the resistance $R_{x,l}$, for nodes that are neighbors in the x-coordinate and $R_{y,l}$ for neighbors in the y-coordinate. Nodes of different layers are vertically connected through the resistances $R_v(i_0, i_1)$ with $i_0 \in L_0$ and $i_1 \in L_1$ with $p(i_0) = p(i_1)$. The nodes i_1 in layer 1 are connected to a heat sink of temperature τ_s with the resistances $R_s(i_1)$ while the nodes i_0 in layer 0 are connected to heat sources $I(i_0)$. Figure 6.7 illustrates the model layout.

In this model the heat flow $I_n(i)$ to a node i from its neighbors $N(i)$ is given as

$$I_n(i) = \sum_{j \in N(i)} \frac{\tau(j) - \tau(i)}{R(i,j)} \qquad (6.5)$$

where $\tau(i)$ is the current temperature of node i and $R(i,j)$ is the resistance between the nodes i and j. The flow I_{sink} to the heat sink is:

$$I_{sink}(i) = \frac{\tau_s - \tau(i)}{R_s(i)} \tag{6.6}$$

For small time intervals Δt the temperature change $\Delta \tau$ of a node i can be approximated as

$$\Delta \tau_i = \frac{I_n(i) + I_{source}(i) + I_{sink}(i)}{C(i)} \Delta t \tag{6.7}$$

where $I_{source}(i)$ is the heat input generated by circuits on the FPGA. Only layer L_0 is connected to sources and only layer L_1 is connected to sinks. Therefore, $I_{source}(i_1) = 0$ and $I_{sink}(i_0) = 0$ for all $i_0 \in L_0$ and $i_1 \in L_1$.

Using a two layer model is a compromise between prediction accuracy and computational efficiency: According to our experiments, one layer model cannot generate the temporal temperature difference we observed while three or more layers do only marginally improve predictions and at the same time generate a greater workload for the learning algorithm because of the increased number of free parameters. The free parameters are the resistors $R(i,j)$, $R_s(i)$, the capacitor $C(i)$, the heat sources $I_{\text{source}}(i)$, $I_{\text{sink}}(i)$ for each node i and, finally, the temperature of the heat sink $\tau_{s,}$.

6.3.1 Learning Thermal Models at Run-Time

The model we propose contains a set P of free parameters that have to be learned by the system in order to make useful predictions. In order to evaluate how good a parameter set P is, we first measure a time series of temperature distributions, then we run a simulation of our model using P and compare the resulting temperature distribution $\tau_s(P, t_i, j)$ at each time step t_i and at each node $j \in L_0$ with the measurement $\tau_m(t_i, j)$. The goal is to find a parameter set P that minimizes the mean square error mse of the simulation given by:

$$mse(P) = \frac{1}{|N||M|} \sum_{i \in M} \sum_{j \in N} (\tau_s(P, t_i, j) - \tau_m(t_i, j))^2 \tag{6.8}$$

where N is the set of layer-0 nodes and M is the set of time indices at which measurements were taken. The learning phase is divided into two stages: First, the system generates a spatially uniform temperature increase by activating all heaters at the same time, see Figure 6.2. Since this leads to a spatially uniform temperature distribution, there is no heat flow I_n between neighboring nodes on the same layer. This allows the system to learn the parameters that determine the vertical heat flow through our model independent of the lateral components. Second, the system generates spatial temperature differences by activating different subsets of the heaters. In the second stage the lateral parameters are learned.

The learning itself is done by randomized hill climbing. Randomized hill climbing [115] is known in the literature as a learning technique that searches for an optimum by moving a state from a single randomly selected starting point towards the direction of the next optimum. For this, the algorithm moves its state in a random direction and evaluates if the state is then closer to a local/global optimum using an objective function. Only if this is the case, the current state is updated by the algorithm. Using this approach the quality of a solution cannot decrease over time. The step size for moving the state decreases over time. Figure 6.8 depicts an example for a 1-dimensional scenario.

The algorithm is called hill climbing because it looks like the state climbs up the next hill, where the objective function forms the landscape and the local/global optima form the top of the hills. The algorithm stops when an optimum is reached or a predefined maximum number of iterations is reached. Since this algorithm tends to reach local optima, several random starting points are usually used.

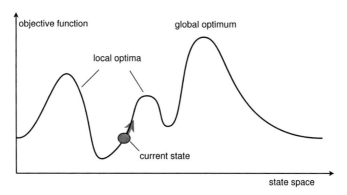

Figure 6.8: Randomized hill climbing: The current state moved towards the next local optimum.

In our implementation, we treat each parameter in the set P as a dimension. The algorithm starts from an initial solution $P = P_{init}$ and through random variation of P tries to maximize the objective function $f = \frac{1}{mse(P)}$. Over a pre-defined number of iterations the random variations become smaller until the algorithm terminates in or close to a maximum.

Source code 6.1 lists the C implementation of our randomized hill climbing algorithm. The algorithm gets an initial parameter set P as input and uses this for a starting point. The algorithm starts with a relative step size $\sigma = \sigma_1$ and decreases this value by a factor of 0.5 until it reaches the minimum relative step size $\sigma = \sigma_2$. Hence, the algorithm takes large steps to climb the hill in the beginning and then finally takes tiny steps at the top of the hill to find the optimum.

The parameter set is altered for each parameter relatively to its old value, see line 10. A Gaussian function is used for randomization of the step size and the direction. The new parameter set $P2$ is evaluated using the objective function f (line 12). If the new parameter set $P2$ is an improvement to the currently best parameter set P, it is updated (lines 13–17). For each step size σ there is a maximum number of iterations (max_iterations). To reduce the average run-time of the algorithm, the number of iterations of each relative step size can be cut short, if there are no improvements of the parameter set over a longer period (max_stagnations iterations). In a next step, the relative step size σ is halved (line 24). Since σ is halved each time, there are $\lceil \log_2(\frac{\sigma_1}{\sigma_2}) \rceil$ different relative step sizes.

To further reduce the run-time of the learning phase, we have divided the algorithm into two stages. For the first stage of the learning phase no lateral heat flow is considered. Therefore, the parameters for each tile can be learned individually with a reduced thermal model that only focuses on one tile. Hence, instead of a 2-layer model with $\frac{N}{2}$ tiles, there are $\frac{N}{2}$ models with a single tile (and two nodes) only. This reduces the computational complexity of computing the objective function from $O(N * M)$ to $O(M)$. Thus, the computational complexity of learning the parameters of one tile is $O(\lceil \log(\frac{\sigma_1}{\sigma_2}) \rceil * \theta * |P_y| * M)$ where θ represents the maximum number of iterations and P_y is the parameter set that is learned in the first stage. P_y contains $C(i_0)$, $C(i_1)$, $R(i_0, i_1)$, $R_s(i_1)$, $I_{source,heater\_off}(i_0)$, $I_{source,heater\_on}(i_0)$ and $I_{sink}(i_1)$ where i_0 and i_1 are the two remaining nodes in the reduced thermal model that covers only a single tile. Since the parameters for $\frac{N}{2}$ tiles have to be computed, the overall complexity of the first stage is $O(\lceil \log(\frac{\sigma_1}{\sigma_2}) \rceil * \theta * N * |P_y| * M)$.

In the second stage, the system generates spatial temperature differences between tiles by selectively activating the heaters to the top, bottom, left and to the right. Since the lateral heat flow is of special interest in this stage, the thermal model cannot be reduced. Hence, the computational complexity of the objective function is $O(N * M)$. This is mainly used to find values for the lateral resistances $R_{x,l}$ and $R_{y,l}$. In our model, we assume that the parameters $R_{x,l}$ and $R_{y,l}$ are the same

```
1   float randomized_hill_climbing(float (*f)(float*,int), float * P,
2       int d, float sigma_1, float sigma_2, int max_iterations,
3       int max_stagnations){
4           float sigma = sigma_1;
5           float P2[d];
6           float f1, f0 = f(P,d);
7           while(sigma >= sigma_2){
8               int i = 0, k = 0, j;
9               while(i<max_iterations && k<max_stagnations){
10                  for(j = 0; j < d; j++){
11                      P2[j] = P[j] + P[j]*sigma*noise();
12                  }
13                  f1 = f(P2,d);
14                  if(f1 > f0){
15                      f0 = f1; k = 0;
16                      for(j = 0; j < d; j++){
17                          P[j] = P2[j];
18                      }
19
20                  } else {
21                      k++;
22                  }
23                  i++;
24              }
25              sigma *= 0.5;
26          }
27          return f0;
28  }
```

Source code 6.1: C source code for randomized hill climbing

between all neighboring tiles of the same layer. The computation complexity of the second stage is $O(\lceil\log(\frac{\sigma_1}{\sigma_2})\rceil * \theta * |P_x| * N * M)$ where $P_x = \{R_{x,l}, R_{y,l} \mid l \in \{0,1\}\}$.

Note that the number of measurement points for the first and second stage of the learning phase might differ since the first stage requires measurements of generating globally uniform heat whereas the second stage requires measurements of generating spatial differences. The overall computational complexity for the learning phase is $O((|P_y| * M_1 + |P_x| * M_2) * \lceil\log(\frac{\sigma_1}{\sigma_2})\rceil * \theta * N)$ where M_1 and M_2 represent the different numbers of measurement points for both stages.

Other learning algorithms such as simulated annealing or evolutionary algorithms can also be applied to learn the model parameters. Evolutionary algorithms do not rely on good initial parameters and often find adequate global solutions. However, the number of parameters to be learned is high, i.e., a thermal model that contains κ tiles has $7 * \kappa + 4$ free parameters. We have observed that by applying evolutionary algorithms as learning strategy resulted in run-times of several days in order to learn the parameters of a thermal model with 10×15 tiles when the employed processor runs at 100 MHz. In contrast, applying hill climbing on the same data reduced the learning time to about an hour with comparable results. In our experimental results, see Section 7.3.4, randomized hill climbing performed well for learning the thermal model of an FPGA. Therefore, we did not further investigate other learning techniques.

6.3.2 Predicting Temperature Distributions

Using the previously learned thermal model, the system becomes temperature-aware and can predict future temperature distributions at run-time: The system initializes the tile temperatures of layer 0 with the current measurements of the temperature sensors. If some of the tiles do not contain temperature sensors, their temperatures can be gauged by neighboring tiles that contain sensors. The temperatures of the tiles in layer 1 can then be initialized so that the system is in thermal equilibrium. For temperature prediction, the changes in temperature have to be updated iteratively for all tiles using Equation 6.7. Continuous updates of the tile temperatures for small Δt values lead to accurate temperature predictions of arbitrary length.

In the case of a self-adaptive hybrid multi-core, each thread may be mapped to a CPU or to a digital circuit implemented on the FPGA. Each thread has a thermal footprint in the form of heat sources $I_{\text{source}}(i)$ that depends on where it is running and on its current workload. These thermal footprints have to be known in advance in order to predict the system's future temperature distribution.

In the case of thread-to-core re-mapping, the heat sources $I_{\text{source}}(i)$ of the affected tiles change. Thus, if the system wants to predict the effect of mapping a thread from one part of the chip to another part, only the heat sources need to be updated. Then, the system can predict the temperature distribution after remapping without the need to actually remap the threads. This is only true, if the system knows the heat sources $I_{\text{source}}(i)$ of the threads. Figure 6.9 depicts an example, where the thermal effects of a thread migration between two cores can be predicted by only altering the heat sources of the corresponding core tiles.

We propose that the temperature-aware system learns the $I_{\text{source}}(i)$ parameters of different threads at run-time and during thread execution. In the beginning, this implies that the system has to explore the effects of different thread mappings. Note that we do not need to learn all possible combinations of thread-to-core mappings. For instance, if the system has $|C|$ cores and $|T|$ threads, $|T|$ measurements are sufficient. Here, the system may map the i-th thread to all $|C|$ cores

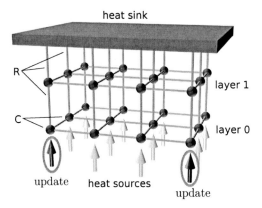

Figure 6.9: Prediction using thermal model: If a thread (represented by its heat source) is migrated from core C_1 (first row, right) to C_2 (first row, left) only the corresponding heat sources need to be updated.

in the i-th measurement. Nevertheless, the system must map each thread to each possible core once to be able to predict accurate temperature distributions.

The temperature predictions can be pruned, when the system reaches a steady state temperature. If an entire thread schedule should be predicted, the system has to change the heat sources of affected tiles for the specific points in time. However, learning the heat sources at run-time is omitted in this thesis. For our self-adaptation algorithms the multi-core system intends to balance the on-chip temperature distribution. Therefore, it uses various heuristics with different knowledge basis. Some heuristics include knowledge about the heat sources of each possible thread-to-core mapping. In this thesis it is assumed, that the system has already learned these parameters at run-time. Detailed temperature predictions, however, were not taken into account to guide the self-adaptation.

6.4 Self-adaptation Algorithms

This section describes different self-adaptation strategies where the thread-to-core mapping is adjusted to balance the on-chip temperature distribution. Thread-to-core mapping combines initial assignment of threads to cores and in the cooperative case also thread migration between cores.

The set of (runnable) threads is defined as $T = \{T_1, \ldots, T_{|T|}\}$ and the set of cores is denoted as $C = \{C_1, \ldots, C_{|C|}\}$. The algorithms map threads to cores. If there are more cores than threads, some cores idle. This possibility is modeled by the extended thread set $T^+ = T \cup \{T_{\text{idle}}\}$. A thread-to-core mapping $\mu : T^+ \mapsto C$ defines for each active thread on which core it runs. Here, each thread can be executed on at most one core except the idle thread, hence $\forall C_j, C_{k \neq j} \in C : (\mu^{-1}(C_j) = \mu^{-1}(C_k)) \rightarrow (\mu^{-1}(C_j) = \mu^{-1}(C_k) = T_{\text{idle}})$.

The self-adaptation techniques do not reject or stop any thread regardless of the measured temperatures of the multi-core. Instead, each thread is executed until completion. When a thread has no more workload, the core either becomes idle or gets another thread assigned. The mapping can be compared to a dispatcher, which decides for a set of runnable threads which thread should be mapped to which core.

The self-adaptation techniques are in charge of thermally balancing the chip by assigning the threads to cores, which either aim at minimizing the thermal imbalance or which increase the thermal imbalance as little as possible. Using thread migration the cooperative mapping algorithms can move local hot spots into colder regions of the chip. More advanced scheduling techniques which perform time slicing and possibly pause threads for selected time periods to regulate the on-chip temperature are out of scope of this thesis.

As defined in Section 3.4.1, the FPGA is partitioned into a regular grid of tiles $\Lambda = \{\Lambda_1, \ldots, \Lambda_{|\Lambda|}\}$. Each tile Λ_l contains a temperature sensor such that the temperature can be accessed for each tile using $\tau : \Lambda \mapsto \mathbb{R}$. All cores stretch across disjunctive sets of tiles. The core map is given by $\lambda : C \mapsto \mathcal{P}(\Lambda)$, where $\mathcal{P}(\Lambda)$ defines the power set of Λ.

Each tile is associated to at most one core $(\forall C_j, C_{k \neq j} \in C : (\lambda(C_j) \cap \lambda(C_k)) = \emptyset)$ and at least one tile is assigned to each core $(\forall C_j \in C : |\lambda(C_j)| > 0)$. When a core executes a thread, the thread imposes heat sources on the tiles of the core, which is given by $\phi : T^+ \times C \times \Lambda \mapsto \mathbb{R}$.

The thread does only create heat sources on the tiles of the corresponding core, the other tiles are not affected directly by this mapping. Therefore, $\forall T_i \in T^+, C_j \in C, \Lambda_l \notin \lambda(C_j) : \phi(T_i, C_j, \Lambda_l) \rightarrow 0$. Note that the heat flow to the other tiles of the chip is modeled by the temperature model (cf. Section 6.3).

In the following Section 6.4.1 non-cooperative thread-to-core mappings, which do not allow for thread migration during thread execution, are discussed. Furthermore, Section 6.4.2 presents cooperative mapping strategies that allow for thread migrations. The different non-cooperative and cooperative mapping strategies will be evaluated using the temperature simulator, presented in Section 6.5. The simulation results can be found in Section 7.3.5.

6.4.1 Non-cooperative Thread-to-core Mapping

For non-cooperative thread-to-core mapping, each thread, which is not already being executed on a core, needs to be assigned to a core as long as there is an idle core left. A thread, which has not completely processed its current input data, cannot be preempted and, therefore, cannot be migrated to another core. The generic non-cooperative thread-to-core mapping algorithm is given in Algorithm 6.1. The algorithm is called (i) when a thread completes its workload and there are waiting runnable threads or (ii) when a thread receives new workload and becomes runnable while there is at least one idle core. In the other cases, the thread-to-core mapping will not be altered. Hence, when there is a runnable thread T_i, which has not been assigned to a core yet (line 2), the algorithms selects from the set of idle cores C' (line 4) the most appropriate core according to a non-cooperative (n) heuristic h_n (line 6).

Algorithm 6.1 Generic non-cooperative thread-to-core mapping algorithm

---

Require: set T of runnable threads, set C of cores, set of tiles $\Lambda = \{\Lambda_1, \dots, \Lambda_{|\Lambda|}\}$, tiles per core $\lambda : C \mapsto \mathcal{P}(\Lambda)$, thread-to-core mapping $\mu : T^+ \mapsto C$, heat sources of affected tiles for each (thread,core) tuple $\phi : T^+ \times C \times \Lambda \mapsto \mathbb{R}$, sensor temperature readings for each tile $\tau : \Lambda \mapsto \mathbb{R}$.

1: **procedure** GENERIC_NON_COOPERATIVE_MAPPING_ALGORITHM
2:    **for all** $T_i \in T$ **do**   ▷ For all runnable threads
3:       **if** $\mu(T_i) \mapsto \emptyset$ **then**   ▷ If thread T_i is not assigned to any core
4:          $C' \leftarrow \{C_j | C_j \in C, \mu^{-1}(C_j) = \emptyset\}$ ▷ Define set of idle cores
5:          **if** $|C'| > 0$ **then**   ▷ If there are idle cores
6:             $C_j \leftarrow h_n(T_i, C', T^+, \Lambda, \lambda(\Lambda), \phi(T^+, C, \Lambda), \tau(\Lambda))$ ▷ Select core
7:             update: $\mu^{-1}(C_j) \mapsto T_i$ ▷ Assign thread to core
8:          **end if**
9:       **end if**
10:   **end for**
11: **end procedure**

This section presents eleven non-cooperative mapping heuristics. Two of these heuristics do not take temperature into account. These heuristics either map the threads to cores naively or randomly ($h_{n,\text{naive/random}}$). In contrast to that, three further heuristics map according to the current temperature sensor readings of the idle cores ($h_{n,\alpha,\text{max/avg/sum}}$).

The next three heuristics, $h_{n,\beta,\text{max/avg/sum}}$, map the threads to cores according to the (learned) heat source values of each (thread, core) tuple. These heuristics assume that the system has learned these parameters at run-time and can now exploit this knowledge for possibly better mapping decisions. The final three non-cooperative mapping strategies, $h_{n,\gamma,\text{max/avg/sum}}$ combine the knowledge of the current temperature distribution, which is given by the temperature sensors, and the learned information about the varying heat sources for different thread-to-core mappings. A list of all non-cooperative mapping heuristics is given in Equation 6.9.

$$
h_n(T_i, C', \ldots, \tau(\Lambda)) = \begin{cases}
h_{n,\text{naive}}(T_i, C', \ldots, \tau(\Lambda)) \\
h_{n,\text{random}}(T_i, C', \ldots, \tau(\Lambda)) \\
h_{n,\alpha,\text{max}}(T_i, C', \ldots, \tau(\Lambda)) \\
h_{n,\alpha,\text{avg}}(T_i, C', \ldots, \tau(\Lambda)) \\
h_{n,\alpha,\text{sum}}(T_i, C', \ldots, \tau(\Lambda)) \\
h_{n,\beta,\text{max}}(T_i, C', \ldots, \tau(\Lambda)) \\
h_{n,\beta,\text{avg}}(T_i, C', \ldots, \tau(\Lambda)) \\
h_{n,\beta,\text{sum}}(T_i, C', \ldots, \tau(\Lambda)) \\
h_{n,\gamma,\text{max}}(T_i, C', \ldots, \tau(\Lambda)) \\
h_{n,\gamma,\text{avg}}(T_i, C', \ldots, \tau(\Lambda)) \\
h_{n,\gamma,\text{sum}}(T_i, C', \ldots, \tau(\Lambda))
\end{cases} \tag{6.9}
$$

The uninformed naive mapping strategy does not intend to minimize the thermal chip balance. This heuristic was developed to compare the effects of introducing thermal management into a (hybrid) multi-core system. If there are more idle cores than runnable threads, the naive mapping heuristic assigns the threads to cores with the smallest identifiers, see Equation 6.10.

$$
h_{n,\text{naive}} = C_{\min\{j \mid C_j \in C'\}} \tag{6.10}
$$

Since the naive mapping tends to generate clusters of active (hot) cores, a second random strategy was implemented, which selects the cores randomly, see Equation 6.11.

$$
h_{n,\text{random}} = \text{random\_choice}(C') \tag{6.11}
$$

The following heuristics take more information such as the sensor input and/or the learned heat sources for all (thread,core) tuples into account. Each core in C' is evaluated using the heuristics $h_{n,\alpha/\beta/\gamma,\text{max/avg/sum}}$. If there are two or more cores with the same evaluation result, a core is selected randomly among these cores.

The heuristics $h_{n,\alpha,\text{max/avg/sum}}$ select the idle core(s) with the coldest temperature(s). Since each core can have multiple sensors, three different criteria for evaluating the core temperature were used: max, sum and avg. For the max criterion, the core temperature is taken to be the temperature of the hottest tile from among all core tiles. The avg criterion assigns the average tile temperature to each core. The max and the avg criteria, do not take the different area sizes of the core into account.

For the sum criterion $h_{n,\alpha,\text{sum}}$, the tile temperatures are adapted in such a way that the overall minimum tile temperature is subtracted from all tile temperatures $\tau(\Lambda_l)$. The resulting tile temperature differences are denoted as $\Delta\tau(\Lambda_l) = \tau(\Lambda_l) - \min_{\Lambda_{l_2} \in \Lambda} \{\tau(\Lambda_{l_2})\}$. The sum criterion $h_{n,\alpha,\text{sum}}$ assigns the sum of the corresponding tile temperature differences to each core C_j. Hence, the sum criterion, favors cores with less tiles. The mapping heuristics for all three criteria can be found in Equations 6.12-6.14. The temperature differences of each tile $\Delta\tau(\Lambda)$ were used because adding the temperature readings $\tau(\Lambda)$ of each tile for all cores disproportionally penalizes large cores.

$$h_{n,\alpha,\text{max}} = \underset{C_j \in C'}{\arg\min} \left\{ \max_{\Lambda_l \in \lambda(C_j)} \{\tau(\Lambda_l)\} \right\} \tag{6.12}$$

$$h_{n,\alpha,\text{avg}} = \underset{C_j \in C'}{\arg\min} \left\{ \frac{\sum\limits_{\Lambda_l \in \lambda(C_j)} \tau(\Lambda_l)}{|\lambda(C_j)|} \right\} \tag{6.13}$$

$$h_{n,\alpha,\text{sum}} = \underset{C_j \in C'}{\arg\min} \left\{ \sum_{\Lambda_l \in \lambda(C_j)} \Delta\tau(\Lambda_l) \right\} \tag{6.14}$$

The following non-cooperative mapping heuristics assume that the system has learned the individual heat source values for all possible thread-to-core mapping. The heuristics $h_{n,\beta,\text{max/avg/sum}}$ select the best possible mappings according to the max, sum and avg criteria. For the sum criterion, the heat sources of considered (thread, core) tuple values are reduced by the heat source of the idle thread. Hence, the sum criterion only focuses on the additional heat source. The heat source-driven mapping heuristics are defined in Equations 6.15-6.17, where $\Delta\phi(T_i, C_j, \Lambda_l) = \phi(T_i, C_j, \Lambda_l) - \phi(T_{\text{idle}}, C_j, \Lambda_l)$. The other criteria are not affected by the differences, because for the min and avg criteria, $\phi(T_{\text{idle}}, C_j, \Lambda_l)$ can be considered a static offset that does not influence the mapping decision in any way.

$$h_{\mathrm{n},\beta,\mathrm{max}} = \underset{C_j \in C'}{\mathrm{argmin}} \left\{ \max_{\Lambda_l \in \lambda(C_j)} \left\{ \phi(T_i, C_j, \Lambda_l) \right\} \right\} \tag{6.15}$$

$$h_{\mathrm{n},\beta,\mathrm{avg}} = \underset{C_j \in C'}{\mathrm{argmin}} \left\{ \frac{\displaystyle\sum_{\Lambda_l \in \lambda(C_j)} \phi(T_i, C_j, \Lambda_l)}{|\lambda(C_j)|} \right\} \tag{6.16}$$

$$h_{\mathrm{n},\beta,\mathrm{sum}} = \underset{C_j \in C'}{\mathrm{argmin}} \left\{ \sum_{\Lambda_l \in \lambda(C_j)} \Delta\phi(T_i, C_j, \Lambda_l) \right\} \tag{6.17}$$

Finally, the hybrid mapping heuristics $h_{\mathrm{n},\gamma,\mathrm{max/avg/sum}}$ combine knowledge of the current temperature distribution of the chip and the heat sources of all (thread,core) tuples, see Equations 6.18-6.20. Therefore, the maximum, the average, and the sum of the term $\Delta\phi(T_i, C_j, \Lambda_l) * \Delta\tau(\Lambda_l)$ is defined for all $\Lambda_l \in \lambda(C_j)$ with $C_j \in C'$.

$$h_{\mathrm{n},\gamma,\mathrm{max}} = \underset{C_j \in C'}{\mathrm{argmin}} \left\{ \max_{\Lambda_l \in \lambda(C_j)} \left\{ \Delta\phi(T_i, C_j, \Lambda_l) * \Delta\tau(\Lambda_l) \right\} \right\} \tag{6.18}$$

$$h_{\mathrm{n},\gamma,\mathrm{avg}} = \underset{C_j \in C'}{\mathrm{argmin}} \left\{ \frac{\displaystyle\sum_{\Lambda_l \in \lambda(C_j)} \Delta\phi(T_i, C_j, \Lambda_l) * \Delta\tau(\Lambda_l)}{|\lambda(C_j)|} \right\} \tag{6.19}$$

$$h_{\mathrm{n},\gamma,\mathrm{sum}} = \underset{C_j \in C'}{\mathrm{argmin}} \left\{ \sum_{\Lambda_l \in \lambda(C_j)} \Delta\phi(T_i, C_j, \Lambda_l) * \Delta\tau(\Lambda_l) \right\} \tag{6.20}$$

6.4.2 Cooperative Thread-to-core Mapping

In contrast to non-cooperative mapping strategies, cooperative mapping strategies allow thread migration between cores during execution under certain conditions. To respect the limited interruptibility of cooperative hybrid threads, which can be executed on hardware and software cores, each thread will be only allowed to migrate once in a defined time period Δt_p, e.g. $\Delta t_p = 1$ second. This means that for each time interval Δt_p a migration may be performed. To allow for trans-modal thread migration is a research topic on its own, which was described in Section 4.9 for the hybrid multi-core. If the threads implement parts of a streaming application, it might be realistic to allow for thread migrations at defined time intervals.

A good example application for such a streaming application is the video object tracker, which has been introduced Section 5.3.3. This application is partitioned into several stages, which process video frames at a given frame rate. Thus, the thread instances per stage can be migrated between cores of different modality in-between frames, where the thread's context is minimal. In our experimental results for performance management, which have been evaluated using a video object tracking application, trans-modal thread migrations were emulated by simple thread terminations and thread creations on the affected cores.

For cooperative mapping, the initial assignment of cores to runnable threads can be made by the heuristics developed for the non-cooperative mapping h_n that were defined in the previous section. Hence, this section focuses on migration heuristics for cooperative (c) mapping h_c. However, since thread migrations impose a significant overhead in time and possibly temperature, the number of such migrations should be limited. Therefore, ten out of eleven cooperative mapping strategies can migrate threads between at most two cores inside the time interval Δt_p. Hence, the numbers of migrated threads in Δt_p is at most two, which is the case when both cores execute a thread. There is only a single thread migration technique that is allowed to perform up to $|C|$ thread migrations in Δt_p. That migration technique rotates the entire set of running threads through the set of cores once in Δt_p.

The rotating thread migration strategy is presented in Algorithm 6.2. All running threads T are migrated from their current cores to neighboring cores in C each Δt_p, with an exception for the first and last core. More specifically, if a thread T_i was running on core C_j, it will be migrated to core C_{j-1}, when $j > 1$ and to $C_{|C|}$ when $j = 1$. This algorithm does not observe any temperature measurements, but can still balance the temperature distribution, because it shifts all local temperature hot spots through all cores. This strategy prevents a situation, where a subset of cores generate local hot spots, which increase the thermal imbalance over time. However, this algorithm comes at a high cost of up to $|C|$ thread migrations each Δt_p when no core is idle.

A second strategy that does not rely on temperature measurements is the random thread migration algorithm, which is presented in Algorithm 6.3. This time a running thread T_i is selected randomly and migrated from its current core C_j to a randomly selected core $C_{k \neq j}$. If a thread is currently executed on the target core, it will be migrated (swapped) to C_j. Compared to the rotating thread migration algorithm, the number of thread migrations in Δt_p is limited to two. However, local hot spots might only be migrated infrequently.

The following heuristics base their migration decision on temperature monitoring and/or learned system models - such as heat sources of all (thread, core) tuples. The generic thread migration algorithm is presented in Algorithm 6.4.

Algorithm 6.2 Rotating thread migration algorithm

Require: set T of runnable threads, set C of cores, thread-to-core mapping $\mu : T^+ \mapsto C$.

1: **procedure** ROTATING_MIGRATION_ALGORITHM
2:     $T_i \leftarrow \mu^{-1}(C_1)$ ▷ Store thread T_i of first core C_1
3:     **for** $j \in \{2, \ldots, |C|\}$ **do**   ▷ For all cores except C_1
4:        $T_l \leftarrow \mu^{-1}(C_j)$ ▷ Get thread T_l from previous core
5:        update: $T_l \mapsto \mu^{-1}(C_{j-1})$ ▷ Assign thread T_l to core C_j
6:     **end for**
7:     update: $T_i \mapsto \mu^{-1}(C_{|C|})$ ▷ Assign thread T_i to last core $C_{|C|}$
8: **end procedure**

Algorithm 6.3 Random thread migration algorithm

Require: set T of runnable threads, set C of cores, thread-to-core mapping $\mu : T^+ \mapsto C$.

1: **procedure** RANDOM_MIGRATION_ALGORITHM
2:     $C'' \leftarrow \{C_j \mid C_j \in C, \mu^{-1}(C_j) \neq T_{\text{idle}}\}$ ▷ Define set of active cores
3:     $C_j \leftarrow \text{random\_choice}(C'')$ ▷ Get random active core
4:     $C_k \leftarrow \text{random\_choice}(C \setminus \{C_j\})$ ▷ Get random target core
5:     $T_i \leftarrow \mu^{-1}(C_j)$
6:     $T_l \leftarrow \mu^{-1}(C_k)$
7:     update: $T_l \mapsto \mu^{-1}(C_j)$ ▷ migrate thread T_l to core C_j
8:     update: $T_i \mapsto \mu^{-1}(C_k)$ ▷ migrate thread T_i to core C_k
9: **end procedure**

Algorithm 6.4 Generic thread migration algorithm

Require: set T of runnable threads, set C of cores, set of tiles $\Lambda = \{\Lambda_1, \ldots, \Lambda_{|\Lambda|}\}$, tiles per core $\lambda : C \mapsto \mathcal{P}(\Lambda)$, thread-to-core mapping $\mu : T^+ \mapsto C$, heat sources of affected tiles for each (thread,core) tuple $\phi : T^+ \times C \times \Lambda \mapsto \mathbb{R}$, sensor temperature readings for each tile $\tau : \Lambda \mapsto \mathbb{R}$.

1: **procedure** GENERIC_COOPERATIVE_MIGRATION_ALGORITHM
2:     $C_j \leftarrow h_\sigma(C, \Lambda, \lambda(C))$ ▷ Get hottest core according to sensors
3:     $T_i \leftarrow \mu^{-1}(C_j)$ ▷ Get corresponding hottest thread T_i
4:     $C_k \leftarrow h_c(T_i, C_j, T, C, \Lambda, \mu(T), \phi(T^+, C, \Lambda), \tau(\Lambda))$ ▷ Select (best) core
5:     **if** $\mu^{-1}(C_j) \neq \mu^{-1}(C_k)$ **then**   ▷ Swap threads between cores
6:        $T_l \leftarrow \mu^{-1}(C_k)$
7:        update: $T_l \mapsto \mu^{-1}(C_j)$ ▷ migrate thread T_l to core C_j
8:        update: $T_i \mapsto \mu^{-1}(C_k)$ ▷ migrate thread T_i to core C_k
9:     **end if**
10: **end procedure**

First, the hottest core is identified using an evaluation heuristic h_σ. For this purpose, three different heuristics $h_{\sigma,\text{max/avg/sum}}$ (line 2) have been defined, which are listed in Equation 6.21. The first heuristic $h_{\sigma,\text{max}}$ selects the core with the highest tile temperature among all cores, the second heuristic $h_{\sigma,\text{avg}}$ selects the highest average tile temperature of all cores, and, finally, the third heuristic $h_{\sigma,\text{sum}}$ adds up the $\Delta\tau$ values of each tile and therefore takes into account the area size of a core. Again, if there are several cores that have the same evaluation result due to h_σ the ties are broken randomly.

$$
h_\sigma(C, \Lambda, \lambda(C)) = \begin{cases} h_{\sigma,\text{max}}(C, \Lambda, \lambda(C)) = \underset{C_j \in C}{\operatorname{argmax}} \left\{ \underset{\Lambda_l \in \lambda(C_j)}{\max} \{\tau(\Lambda_l)\} \right\} \\[3mm] h_{\sigma,\text{avg}}(C, \Lambda, \lambda(C)) = \underset{C_j \in C}{\operatorname{argmax}} \left\{ \dfrac{\sum\limits_{\Lambda_l \in \lambda(C_j)} \tau(\Lambda_l)}{|\lambda(C_j)|} \right\} \\[3mm] h_{\sigma,\text{sum}}(C, \Lambda, \lambda(C)) = \underset{C_j \in C}{\operatorname{argmax}} \left\{ \sum\limits_{\Lambda_l \in \lambda(C_j)} \Delta\tau(\Lambda_l) \right\} \end{cases} \quad (6.21)
$$

When the hottest core C_j has been identified, the thread T_i being executed on C_j is going to be migrated to another core, if the migration heuristic h_c returns a different core than C_j (line 4). Note that no thread needs to be migrated if C_j is idle. However, this case presumably only occurs when all cores are idle. The studied thread migration heuristics are listed in Equation 6.22.

$$
h_c(T_i, C_j, \ldots, \tau(\Lambda)) = \begin{cases} h_{c,\alpha,\text{max}}(T_i, C_j, \ldots, \tau(\Lambda)) \\ h_{c,\alpha,\text{sum}}(T_i, C_j, \ldots, \tau(\Lambda)) \\ h_{c,\alpha,\text{avg}}(T_i, C_j, \ldots, \tau(\Lambda)) \\ h_{c,\beta,\text{max}}(T_i, C_j, \ldots, \tau(\Lambda)) \\ h_{c,\beta,\text{sum}}(T_i, C_j, \ldots, \tau(\Lambda)) \\ h_{c,\beta,\text{avg}}(T_i, C_j, \ldots, \tau(\Lambda)) \\ h_{c,\gamma,\text{max}}(T_i, C_j, \ldots, \tau(\Lambda)) \\ h_{c,\gamma,\text{sum}}(T_i, C_j, \ldots, \tau(\Lambda)) \\ h_{c,\gamma,\text{avg}}(T_i, C_j, \ldots, \tau(\Lambda)) \end{cases} \quad (6.22)
$$

The heuristics $h_{c,\alpha,\text{max/avg/sum}}$ select the coldest core as target for the thread migration. Again the cores are evaluated using the `max`, `avg` and `sum` criteria.

Furthermore, for the sum criterion, the $\Delta\tau(\Lambda)$ were used instead of the $\tau(\Lambda)$ values. The temperature readings on their own would disproportionately favor small cores as targets. The heuristics are defined in Equations 6.23-6.25.

$$h_{c,\alpha,\max} = \underset{C_k \in C}{\operatorname{argmin}} \left\{ \max_{\Lambda_l \in \lambda(C_k)} \{\tau(\Lambda_l)\} \right\} \tag{6.23}$$

$$h_{c,\alpha,\mathrm{sum}} = \underset{C_k \in C}{\operatorname{argmin}} \left\{ \sum_{\Lambda_l \in \lambda(C_k)} \Delta\tau(\Lambda_l) \right\} \tag{6.24}$$

$$h_{c,\alpha,\mathrm{avg}} = \underset{C_k \in C}{\operatorname{argmin}} \left\{ \frac{\sum\limits_{\Lambda_l \in \lambda(C_k)} \tau(\Lambda_l)}{|\lambda(C_k)|} \right\} \tag{6.25}$$

The final heuristics $h_{c,\beta,\max/\mathrm{avg}/\mathrm{sum}}$ and $h_{c,\gamma,\max/\mathrm{avg}/\mathrm{sum}}$ base their migration decision either entirely or partly on the heat sources of different (thread,core) tuples. Therefore, these heuristics first evaluate the current mapping and compare it to the results of each possible thread migration.

More specifically, let C_j be the hottest core according to h_σ and T_i be the thread, which runs on C_j. The mapping heuristics evaluate for each possible target core $C_{k\neq j} \in C$ if the thermal imbalance can be improved by swapping the threads between C_j and C_k. If the target core C_k also executes a thread T_l, this migration should only be performed if the benefits of migrating T_i to C_k are not eliminated by the drawbacks of migrating T_l to C_j. Hence, in the following, a thread T_i is only migrated from C_j to C_k when the mapping $(T_i \mapsto \mu'^{-1}(C_j), T_i \mapsto \mu'^{-1}(C_k))$ is an improvement over the current mapping $(T_i \mapsto \mu^{-1}(C_j), T_l \mapsto \mu^{-1}(C_k))$.

The heuristics $h_{c,\beta,\max/\mathrm{avg}/\mathrm{sum}}$ base their decisions entirely on the heat sources of the corresponding migration variants. To evaluate if a thread migration between C_j and C_k is promising, the heuristics compute the differences between the heat sources of the current mapping and the possible mapping after migration. For this purpose, $\Delta\phi_{T_i}(T_l, C_j, \Lambda_l)$ computes the differences in heat sources between a mapping where T_l is executed on C_j and the mapping where T_i is executed on C_j or more precisely: $\Delta\phi_{T_i}(T_l, C_j, \Lambda_l) = \phi(T_l, C_j, \Lambda_l) - \phi(T_i, C_j, \Lambda_l)$ where $\Lambda_l \in \lambda(C_j)$.

The $\Delta\phi$ values have to be computed for both affected cores C_j and C_k for the thread T_i that should be migrated and the thread $\mu^{-1}(C_k)$ being currently executed on C_k. The max, avg and sum criteria are applied again, as can be seen in Equations 6.26-6.28. Note that when no migration seems promising the heuristics select the $C_k = C_j$ and no thread migration takes place.

$$h_{c,\beta,\max} = \underset{C_k \in C}{\arg\min} \left\{ \max_{\Lambda_l \in \lambda(C_j)} \left\{ \phi(\mu^{-1}(C_k), C_j, \Lambda_l) \right\} - \max_{\Lambda_l \in \lambda(C_j)} \left\{ \phi(T_i, C_j, \Lambda_l) \right\} + \right.$$
$$\left. \max_{\Lambda_l \in \lambda(C_k)} \left\{ \phi(T_i, C_k, \Lambda_l) \right\} - \max_{\Lambda_l \in \lambda(C_k)} \left\{ \phi(\mu^{-1}(C_k), C_k, \Lambda_l) \right\} \right\} \tag{6.26}$$

$$h_{c,\beta,\text{sum}} = \underset{C_k \in C}{\arg\min} \left\{ \sum_{\Lambda_l \in \lambda(C_j)} \Delta\phi_{T_i}(\mu^{-1}(C_k), C_j, \Lambda_l) - \sum_{\Lambda_l \in \lambda(C_k)} \Delta\phi_{T_i}(\mu^{-1}(C_k), C_k, \Lambda_l) \right\} \tag{6.27}$$

$$h_{c,\beta,\text{avg}} = \underset{C_k \in C}{\arg\min} \left\{ \frac{\sum_{\Lambda_l \in \lambda(C_j)} \Delta\phi_{T_i}(\mu^{-1}(C_k), C_j, \Lambda_l)}{|\lambda(C_j)|} - \frac{\sum_{\Lambda_l \in \lambda(C_k)} \Delta\phi_{T_i}(\mu^{-1}(C_k), C_k, \Lambda_l)}{|\lambda(C_k)|} \right\} \tag{6.28}$$

$$h_{c,\gamma,\max} = \underset{C_k \in C}{\arg\min} \left\{ \max_{\Lambda_l \in \lambda(C_j)} \left\{ \Delta\phi(\mu^{-1}(C_k), C_j, \Lambda_l) * \Delta\tau(\Lambda_l) \right\} - \max_{\Lambda_l \in \lambda(C_j)} \left\{ \Delta\phi(T_i, C_j, \Lambda_l) * \Delta\tau(\Lambda_l) \right\} + \right.$$
$$\left. \max_{\Lambda_l \in \lambda(C_k)} \left\{ \Delta\phi(T_i, C_k, \Lambda_l) * \Delta\tau(\Lambda_l) \right\} - \max_{\Lambda_l \in \lambda(C_k)} \left\{ \Delta\phi(\mu^{-1}(C_k), C_k, \Lambda_l) * \Delta\tau(\Lambda_l) \right\} \right\} \tag{6.29}$$

$$h_{c,\gamma,\text{sum}} = \underset{C_k \in C}{\arg\min} \left\{ \sum_{\Lambda_l \in \lambda(C_j)} \Delta\phi_{T_i}(\mu^{-1}(C_k), C_j, \Lambda_l) * \Delta\tau(\Lambda_l) - \sum_{\Lambda_l \in \lambda(C_k)} \Delta\phi_{T_i}(\mu^{-1}(C_k), C_k, \Lambda_l) * \Delta\tau(\Lambda_l) \right\} \tag{6.30}$$

$$h_{c,\gamma,\text{avg}} = \underset{C_k \in C}{\arg\min} \left\{ \frac{\sum_{\Lambda_l \in \lambda(C_j)} (\Delta\phi_{T_i}(\mu^{-1}(C_k), C_k, \Lambda_l) * \Delta\tau(\Lambda_l))}{|\lambda(C_j)|} - \frac{\sum_{\Lambda_l \in \lambda(C_k)} \Delta\phi_{T_i}(\mu^{-1}(C_k), C_k, \Lambda_l) * \Delta\tau(\Lambda_l)}{|\lambda(C_k)|} \right\} \tag{6.31}$$

111

Finally, the hybrid migration heuristics $h_{c,\gamma,\mathrm{max/avg/sum}}$ are defined in Equations 6.29-6.31. Compared to the heuristics $h_{c,\beta,\mathrm{max/avg/sum}}$, the individual thread-to-core mappings for C_j and $C_k \in C$ are not only based on the differences between the corresponding heat sources $\Delta\phi_{T_i}(T_l, C_j, \Lambda_l)$ and $\Delta\phi_{T_i}(T_l, C_k, \Lambda_l)$ but also on the current (reduced) temperature readings for the affected tiles $\Delta\tau(\Lambda_l)$ where $\Lambda_l \in \lambda(C_j) \cup \lambda(C_k)$. This heuristic should prevent that a thread becomes a local hot spot on the FPGA although it is mapped on the most appropriate core. Here, after some time, the temperature becomes so high that a thread migration will eventually become promising, which stands in contrast to the heuristics $h_{c,\beta,\mathrm{max/avg/sum}}$.

6.5 Temperature Simulator

A temperature simulator was used to evaluate different self-adaptation strategies. In our experiments we have learned two different heat sources for inactive and active heaters. These values provide a basis for defining multiple thermal scenarios using varying heat sources, which can be analyzed with our temperature simulator. The underlying thermal model is equivalent to the model defined in Section 6.3.

The actual model parameters have been identified using measurements on a Xilinx Virtex-6 LX240T FPGA by applying a randomized hill climbing learning approach as defined in Section 6.3.1. The detailed results of the learning phase along with the achieved simulation accuracy will be presented in Section 7.3.4. In our experimental results, the root mean square error between simulations and measurements is about $0.72°C$ for a scenario, where the measured temperatures stay inside an interval of $[47°C, 56°C]$, Hence, we believe that our simulation results can be applied to real FPGA systems.

Our simulator takes the FPGA layout, the thermal model parameters, the heat sources of different thread-to-core mappings, a simulation duration, and, either a pre-defined schedule or a workload timetable as input. The workload timetable defines at which time new workload for a specific thread arises. In contrast to a pre-defined schedule, the simulator is in charge of mapping the threads to cores and possibly migrating the threads between different cores during execution. Therefore, the developer can program different mapping strategies into a well-defined function stub, which will be called once each time period Δt_p. At these points, new threads can be assigned to free cores and already active threads can be migrated between cores. Using the function stubs for thread re-mapping, we have implemented the proposed self-adaptation strategies. A detailed evaluation will follow in Section 7.3.5.

The simulator computes the temperatures for each tile on the FPGA each Δt_p and stores this information together with the current thread-to-core mapping into an output file. Most input parameters are given in plain text files. To evaluate a wide range of scenarios these files can be generated automatically with different parameters using dedicated python scripts and shell scripts.

The layout file contains the regular grid size and the core map. The thermal model file defines all static model parameters and the heat sources file stores the heat sources for each (thread, core) tuple of the affected tiles. The workload file lists all time steps of the simulation at which new workload for specific threads arise. Finally, the schedule file contains a list of time steps at which the thread-to-core mapping changes together with the corresponding mapping. The simulator was programmed in C.

6.5.1 Graphical User Interface

The temperature simulation can be visualized using a graphical user interface programmed in Java. To visualize the temperature simulation, the user has to load the FPGA layout and the output file of the simulator. The GUI allows the user to replay the entire simulation at different velocities. The user has also the option to step manually through the entire simulation and to jump to a specific time step.

The tiles are colored according to their current temperature for a better visualization of the temperature distribution. The highest temperature in the simulation is encoded red and the coldest temperature is encoded blue. For the entire temperature scale different compositions between the colors blue-green-yellow-orange-red are used. The detailed temperature values per tile and the core map can be dynamically activated/deactivated. A screenshot of the graphical user interface can be seen in Figure 6.10.

The GUI provides the user with valuable information to evaluate the thermal balance of the FPGA. The maximum and minimum temperature and the difference between both temperatures is displayed for the current time step. The maximum temperature and minimum temperature and the maximal temperature difference in a temperature distribution is also given for the entire simulation. Finally, the maximal spatial temperature difference between two neighboring tiles is given for the current temperature distribution and the entire simulation.

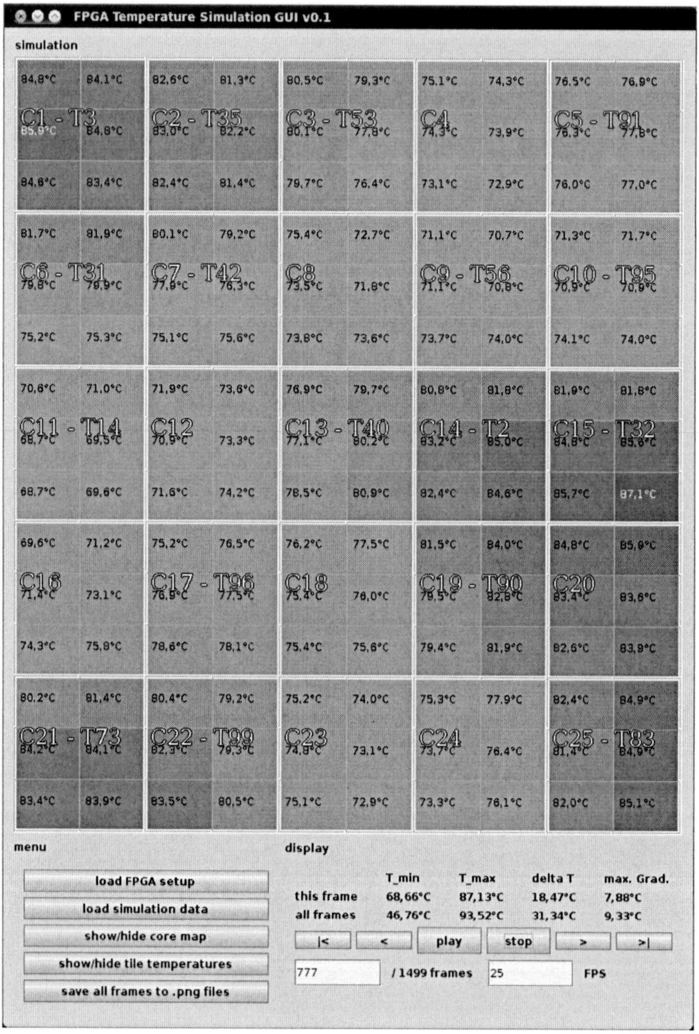

Figure 6.10: Temperature simulator GUI: The simulated hybrid multi-core consists of 25 cores. The FPGA is partitioned into a 10 × 15 regular grid of tiles. The simulated temperature for each tile is displayed.

To compare the temperature simulations of different self-adaptation strategies, another graphical user interface was programmed, which displays two temperature simulations next to each other. Note that for comparing two temperature simulations, they need to have the same FPGA layout and the same input (thread set, workload, duration). The colors of the tiles are calibrated in such a way that the color ranges from the minimum tile temperature of both simulations to the maximum temperature of both simulations.

For instance, if the first self-adaptation strategy is outperformed by the second strategy, the tiles of the second temperature simulation might not be colored red over the entire simulation. This visualizes that the maximum tile temperature of the first simulation is considerably higher than the maximum tile temperature of the second simulation and graphically indicates that the second strategy is superior to the first strategy.

6.6 Chapter Conclusion

In this chapter, a method for measuring the temperature distribution on modern FPGAs was presented. Nowadays the vendors of FPGAs already integrate a thermal diode on the FPGA. However, for detailed thermal management of a hybrid multi-core one sensor node is not sufficient. Therefore, the FPGA is conceptually partitioned into a regular grid of tiles where each tile contains a ring oscillator-based temperature sensor. While ring oscillators are widely used in many academic projects to measure temperature in FPGAs, a novel self-calibration technique has been presented in this chapter.

Traditionally the frequencies of the ring oscillators have been translated to temperatures using extensive manual calibration either using a temperature controlled oven or a thermal infrared camera. Instead of that, the presented novel self-calibration technique uses dedicated heat-generating cores to globally heat up the FPGA. The temperature sensors are then calibrated against the built-in thermal diode. A special focus has been set on the design space exploration of temperature sensors, which supports a designer to identify the best sensor layout.

To demonstrate significant thermal imbalances on todays FPGAs, eight different dedicated heat-generating cores (heaters) have been introduced in this chapter, which have the single purpose of heating up the area that contains the heater. Therefore, the different resource types of the FPGA, such as look-up tables, flip-flops, digital signal processor blocks, and block RAMs have been studied separately and partly in combination. These heaters have been successfully used for sensor calibration. The corresponding experimental results will be presented in the next chapter. Using these heaters, a hybrid multi-core can create different spatial temperature differences by configuring these heaters to different subsets of its cores.

To evaluate the thermal effects of thread migrations, a thermal model has been presented in this chapter. The heat flow between the tiles of the FPGA can then be measured using the sensor grid and the corresponding temperature readings can serve as an input to learn a thermal model of the chip. The proposed two-layered thermal model bases on the multi-layered thermal model of the temperature simulation tool HotSpot [73], which makes use of the well-known duality between thermal and electrical phenomena.

HotSpot derives the model parameters of the material properties of the different layers of the chip. In contrast to HotSpot and most related work, the model parameters are learned at run-time using a learning algorithm based on randomized hill climbing. Thus, a hybrid multi-core system can generate its own thermal model without the knowledge of the material properties. Moreover, the thermal model can be used to predict temperature distributions on the FPGA. For instance, if the thermal footprints of the individual threads are learned at run-time, the model can predict the thermal effects of different thread migrations.

Multiple self-adaptation techniques have been presented that assign new threads to cores and possibly migrate running threads in order to balance the on-chip temperature distribution. Therefore, twenty-two self-adaptation heuristics have been introduced in this chapter. The first half of the heuristics assume that thread migrations are prohibited during thread execution, whereas the second half of the heuristics allow thread migrations. The particular heuristics base their mapping decisions on different levels of knowledge including no knowledge, being able to measure the temperature distribution, and, knowing the thermal footprints of each thread-to-core mapping. A temperature simulator and a corresponding graphical user interface have been presented in this chapter, which are later used to evaluate the individual self-adaptation techniques for self-adaptive hybrid multi-cores.

The next chapter will provide the experimental results and simulation results for the performance management and the thermal management on self-adaptive hybrid multi-cores. This includes the results on the design space exploration of ring oscillator-based temperature sensors, the sensor self-calibration, the heat-generating cores, the on-line learning of the thermal model parameters, the initial temperature prediction results, and, finally, an extensive analysis of all presented self-adaptation techniques for thermal management.

((*In another moment down went Alice after it, never once*
considering how in the world she was to get out again. *))*

Lewis Carroll, *Alice's Adventures In Wonderland*

CHAPTER 7

Experimental Results

This chapter provides experimental results for performance and thermal management on self-adaptive hybrid multi-cores. Therefore, Section 7.1 introduces our
prototype platforms. For performance management, a video object tracker, which
has been introduced in Section 5.3.3, was studied as a real-world streaming case
study for static and self-adaptive mapping. The results for the self-adaptation
techniques will be shown together with results for static hybrid multi-cores in
Section 7.2.

For thermal management, Sections 7.3.1-7.3.2 presents results for temperature
monitoring, heat generation, and sensor self-calibration on current FPGA technologies. Section 7.3.4 discusses our experimental results on the on-line learning
of thermal model parameters. Finally, Section 7.3.5 presents and analyzes simulation results of our self-adaptation techniques, which have been defined in
Section 6.4.

7.1 Prototype Platforms

One of the most popular FPGA series from Xilinx is the Virtex family because
these chips usually contain more programmable logic resources than the other
FPGA families from Xilinx, e.g. the Spartan family [155], which targets small
low-power designs. Over the time line of this research project several new
Virtex FPGAs have been released by the vendor. Three different Virtex FPGAs,
namely a Virtex-4, a Virtex-5, and a Virtex-6 FPGA, were used for experimental

evaluation. The Virtex-4 FPGAs are based on 90 nm designs [156], the Virtex-5 FPGAs on 65 nm designs [154] and the Virtex-6 FPGAs on 40 nm designs [162]. The trend of shrinking device structures for FPGA devices continues. The latest Virtex-7 family is built on a 28 nm process technology [159]. Detailed information about the used FPGA boards are listed in Table 7.1.

Table 7.1: Information about the used Xilinx Virtex FPGAs [154, 156, 162]

FPGA board	ML410	ML509	ML605
device	xc4vfx100	xc5vlx110t	xc6vlx240t
package	ff1152	ff1136	ff1156
speed grade	-11	-1	-1
nm technology	90 nm	65 nm	40 nm
look-up tables	LUT4: 84,352	LUT6: 69,120	LUT6: 150,720
flip-flops	84,352	69,120	301,440
block RAMs (18KB)	376	296	832
DSP blocks	160	64	768
PowerPC blocks	2	-	-
main memory	256 MB	256 MB	512 MB
thermal diodes	-	1	1
cooling elements	heat sink	heat sink	heat sink, fan

Table 7.1 shows that the vendor does not integrate hard-core PowerPC processor blocks into the FPGA chip anymore for the latest FPGA Virtex series. Although there exist Virtex-5 FPGAs that still integrate PowerPC processor blocks, the used ML509 board of the Xilinx university program does not include these processors. Since the Virtex-6 series hard-core PowerPC processor blocks are no longer integrated into any Xilinx FPGAs anymore. Hence, on modern Virtex FPGAs soft-core processors had to be used instead. For this reason Xilinx MicroBlaze processors [161], which are based on a 32-bit RISC Harvard architecture, have been instantiated in the programmable logic of the FPGA in order to replace the hard-core PowerPC in our hybrid multi-core prototypes.

Furthermore, Table 7.1 indicates that the chip's temperature becomes more critical for decreasing device structures. The 90nm Virtex-4 FPGA has only a heat sink without any temperature sensor. In contrast to that, a temperature sensor has been integrated for the 65nm Virtex-5 and 40nm Virtex-6 FPGAs. Moreover, for 40nm Virtex-6 FPGAs the temperature becomes so high that the vendor places an active cooling element, a fan, on top of the heat sink.

Experimental results on performance management were obtained on a ML410 FPGA board, whereas the experimental results on thermal management were obtained on ML509 and ML605 FPGA boards.

7.2 Performance Management

This section presents the experimental results on performance management, which were achieved by applying the thread-based developed self-adaptation techniques ATP_bound and ATP_budget to a video object tracking case study.

7.2.1 Experimental Setup

The experimental evaluation for performance management on self-adaptive hybrid multi-cores have been performed on the ML410 FPGA board, which contains a Virtex-4 FPGA. The FPGA contains two PowerPC processor blocks where one processor is employed as main processor, and the other as worker processor. Both processors are clocked at 300 MHz. Furthermore, the hybrid multi-core contains two reconfigurable hardware cores, which are clocked at 100 MHz. A sketch of the floorplan that shows the areas of the individual cores on the FPGA can be seen in Figure 7.1.

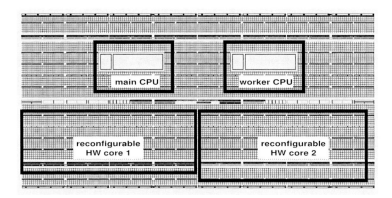

Figure 7.1: Floorplan of the self-adaptive hybrid multi-core for the ML410 FPGA

A particle filter-based video object tracker is used to evaluate the applicability of the developed thread-based self-adaptation techniques in a real-world scenario.

The user streams a video from a workstation, i.e. a personal computer (PC), to the FPGA board using a TCP/IP[22] connection over Ethernet. The raw video frames are sent in the HSV color space to the FPGA. The user can either send the live input of a webcam or, alternatively, a prerecorded video file to the FPGA. Initially the user selects an object in the first frame of the video stream. This information is communicated to the FPGA also over the TCP/IP connection. The experimental setup for the video tracking case study, which implements the particle filter (PF) framework (described in Section 5.3.2), is depicted in Figure 7.2.

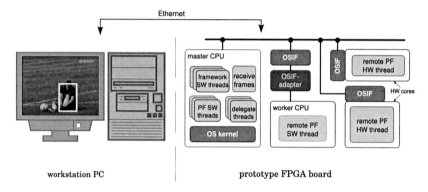

Figure 7.2: Experimental setup for the video object tracking case study

Since the video transfer with raw images over TCP/IP is a bottleneck for the video tracking application, it is assumed that the video frames are available at the hybrid multi-core. To emulate this, a video is transmitted in chunks of frames to the FPGA. When a full chunk has been transferred to the FPGA, the particle filter tracks the object in the chunk of frames. Afterwards it returns the estimation of the object's position and size for the last frame of the chunk back to the workstation. A graphical user interface on the workstation displays the frame and visualizes the estimated position and size of the object by drawing a rectangle around it. Thus, the user can judge intuitively, if the tracking is successful. The performance measurements that are presented in Section 7.2 omit the video transfer of the FPGA board. Note that the chunk size can be reduced to one for a live video tracking using a webcam. To reduce the overhead for the video transfer the video can be downscaled. However, the video was sent in its original size to the FPGA for all experimental results, which are discussed in this thesis.

---

[22]Transmission Control Protocol / Internet Protocol (TCP/IP)

The main goal of our self-adaptive approach is to increase the efficiency with which an application that has been designed using our framework utilizes the reconfigurable area and the worker processors. Regarding the run-time dynamics of our video object tracking case study different thread-to-core mappings with varying area or core requirements may suffice to keep the the tracking performance (measured in frames per second) either above a user-defined performance bound or inside a user-defined performance budget. Therefore, we apply the self-adaptation algorithms ATP_{bound} and ATP_{budget} introduced in Section 5.2 to minimize the required processing resources.

The video object tracker is a prime example to demonstrate self-adaptive thread-to-core mapping for two reasons. First, the required processing power strongly depends on the current contents of the video frames and can vary significantly. Second, the data-parallelism of individual stages of the application favors multithreaded execution where adding more instances of the same thread helps increasing the performance (see Figure 5.3).

7.2.2 Static Mappings

In a first step, static thread-to-core mappings are studied that do not adapt the mapping at run-time. When these static mappings face the same tracking scenario, it can be seen which mapping is superior for each frame of the example video. In this application, placing the sampling (s) or resampling (r) stages in hardware does not yield any performance improvements; for simplicity, these mappings have not been included in the discussion. Thus, the focus is set on the thread-to-core mappings that assign thread instances of the observation (o) and importance (i) stages of the particle filter on the hybrid multi-core.

All studied static thread-to-core mappings are listed in Table 7.2, where $sw_{o/i}$ denotes the thread-to-core mapping for the worker processor and $hw_{o/i/oo/oi/ii}$ represents the thread-to-core mappings for the reconfigurable hardware cores. Moreover, the mapping sw_{i*} shows a mapping where the worker processor executes the importance stage clocked at 100 MHz instead of 300 MHz. This mapping is only available for the static mappings, because the clock frequency cannot be changed at run-time in the presented self-adaptive multi-core prototype, which only allows thread-based adaptations. As the hardware thread implementations are the same for both reconfigurable hardware cores, the tracking performance (in execution time) does not change when a hardware thread is mapped on the first or the second hardware core. Therefore, there was no need to distinguish between different thread-to-core mappings for individual hardware threads for this case study.

Table 7.2: Static mappings for the video object tracking case study

mapping	worker CPU	1st HW core	2nd HW core
sw	-	-	-
hw_i	-	importance	-
hw_o	-	observation	-
sw_i	importance	-	-
sw_{i*}	importance	-	-
sw_o	observation	-	-
hw_{ii}	-	importance	importance
hw_{oo}	-	observation	observation
hw_{oi}	-	observation	importance
sw_i,hw_i	importance	importance	-
sw_i,hw_o	importance	observation	-
sw_{i*},hw_i	importance	importance	-
sw_{i*},hw_o	importance	observation	-
sw_o,hw_i	observation	importance	-
sw_o,hw_o	observation	observation	-
sw_i,hw_{ii}	importance	importance	importance
sw_i,hw_{oo}	importance	observation	observation
sw_i,hw_{oi}	importance	observation	importance
sw_{i*},hw_{ii}	importance	importance	importance
sw_{i*},hw_{oo}	importance	observation	observation
sw_{i*},hw_{oi}	importance	observation	importance
sw_o,hw_{ii}	observation	importance	importance
sw_o,hw_{oo}	observation	observation	observation
sw_o,hw_{oi}	observation	observation	importance

Figure 7.3 depicts the performance figures, measured in clock cycles per frame, of the individual mappings when tracking a soccer player in the video sequence (see Figure 5.5, frame size: 480×360). In all experiments, the particle filter tracks $N = 100$ particles divided into chunks of 10. For visual clarity, Figure 7.3 is split into eight diagrams.

It can be seen that the overall particle filter performance is data-dependent for all mappings. This is mainly due to the histogram creation in the observation stage, as the computational complexity per particle depends on the size of the bounding box. In the first 100 frames, when the player occupies a larger region in the foreground, the filter performance is comparably low. Over the following 100 frames, the performance increases as the player retreats into the background, since smaller bounding boxes and thus fewer pixels need to be observed per frame. In the soccer video, the number of pixels occupied by a particle's bounding box varies over time by a factor of 40, resulting in large variations of the observation stage's execution time.

In contrast, the performance of the importance stage remains constant over the course of the sequence, because the filter compares fixed-size color histograms, which do not depend on the particle's scaling factor. Thus, mapping a single importance thread to a hardware or software worker core benefits the performance across the entire video. The execution time of the remote threads depend on the clocking frequencies of the worker cores. Hence, clocking the worker CPU at 100 MHz instead of 300 MHz results in a triplication of the execution time. The effect of the different frequencies of the worker processor on the overall tracking performance can be seen in the graphs for the mapping sw_i and sw_{i*}.

Furthermore, the difference in the input-data dependency of the observation and importance stages is illustrated in Figure 7.4, which highlights the individual stages' execution times over the video sequence of a software-only mapping (sw). Additional observation threads being executed in parallel improve the performance significantly when the player is in the foreground and the histogram calculation involves a higher number of pixels, i.e., in the first 100 frames of the video. Hence, this interval shows the highest differences between all studied thread-to-core mappings.

The case study also exhibits a non-uniform performance distribution across hybrid worker cores. For the importance stage, the hardware core provides a higher performance gain than the software core when the input-data size stays roughly the same (frames 0–100 and frames 200–300). Interestingly, the remote worker CPU (clocked at 300 MHz) performs better than the hardware core when the input-data size is changing dramatically (frames 100–200), although the importance stage should be data-independent, as it compares only fixed-sized histograms. The particle filter processing includes, by definition, a certain degree of randomness inside the sampling stage that transforms the particles. Thus, for different measurements the bounding boxes of the particles can shrink in a different dynamic over time, which results in different performance behavior over this time period. For each thread-to-core mapping, the results of 50 measurements have been averaged to reduce the impact of statistical variation.

When looking at the differences in performance of individual worker cores, the remote software observation thread generally shows better performance than the hardware thread implementation. However, this difference becomes insignificant when the tracked object's size decreases, and the worker cores cannot significantly reduce the observation stage's workload for the master processor.

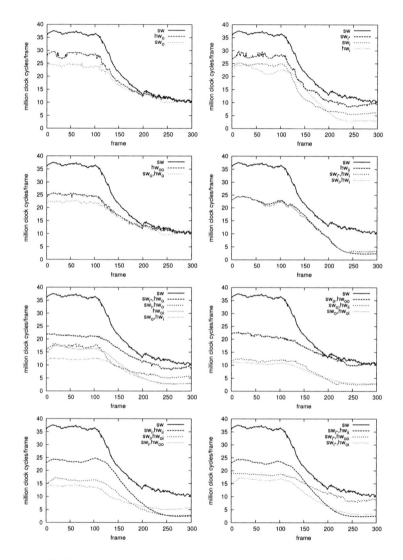

Figure 7.3: Static performance measurements for object tracking in a video sequence (soccer) using one worker core, two worker cores and three worker cores. [8]

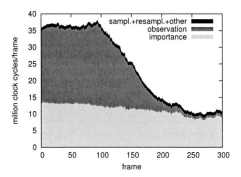

Figure 7.4: Individual execution times of the observation and importance stages for each frame when the application runs entirely on the master. [8]

In general, a combination of both remote software and hardware threads shows the best performance, as it utilizes more computing resources and enables the highest amount of parallel execution. Furthermore, it can be seen that mappings with multiple remote threads for the importance stage do not decrease the execution time compared to mappings with a single hardware thread for the importance stage.

7.2.3 Self-adaptation Strategy with Bound

The static mapping results indicate that the various thread-to-core mappings show different performance curves (in execution time) over time for a given tracking scenario. The execution time strongly depends on the dynamics of the input data. Hence, it is very hard to find the best thread-to-core mapping for a given performance bound or performance interval at design time. The developed self-adaptation strategies aim to reduce the number of used worker cores while either providing a minimal user-defined performance bound or by meeting a user-defined performance interval. The self-adaptation algorithms monitor the execution times of each stage measured on the master processor. When the measured performance does not meet the user-defined requirements, the algorithms exploit their knowledge about the speedup factors of different (thread, core) mappings. The speedup factor $S_{i,j}$ indicates how much faster a thread instance $T_{i,j}$ can be executed on a core C_j compared to the scenario where the thread T_i runs on the master processor instead.

125

Table 7.3: Speedup values $S_{i,j}$ for thread instances $T_{i,j}$. [8]

stage	master CPU	worker CPU	HW core
observation	1.0	1.3	0.6
importance	1.0	1.4	14.4

The time interval for running the self-adaptation algorithm controls a trade-off between the overhead incurred by partial reconfiguration and the latency with which the algorithm reacts to changing data-dependent thread performance. The overhead influences the overall application efficiency, while the reaction latency is relevant for the real-time characterization of the system. In this example, the self-adaptation algorithm is called every 20 frames with an initial offset of eight frames. The speedup values $S_{i,j}$ have been obtained in static measurements and are given in Table 7.3.

The software threads can be executed faster on the worker processors than on the the master processor because the threads cannot be interrupted by other threads of the particle filter framework or by the operating system. The observation stage is slower in hardware than in software, because the software threads internally cache the histogram values for each pixel, which is not done in hardware. Furthermore, the CPUs have a higher memory bandwidth because the experiments have been performed using ReconOS version 2.01, which did not allow stable burst transfers of arbitrary length. Since the observation thread requires a dynamically changing numbers of pixels, the pixel data of the current frame had to be transferred to the hardware core using single-word memory accesses. Despite the low memory bandwidth, the importance thread runs 14.4 times faster on a reconfigurable hardware core than on the master CPU.

Figure 7.5 shows an example run of the system using the ATP_{bound} algorithm (Algorithm 5.1) that uses a single lower performance constraint and removes thread instances on worker cores if the estimated resulting performance is estimated to stay above the user-defined soft real-time constraint. As input, the soccer video (see Figure 5.5) was altered, such that the first 300 frames are played four times (forward, backward, forward, backward). This leads to a video where the player runs into the background and then returns to the foreground, twice, indicated at the top of Figure 7.5.

This example run meets the real-time constraint 86% of the time, excluding the initialization phase within the first 30 frames where first thread instances are created on the worker cores. Although all cores run thread instances of the framework, the constraint is missed in an interval of about 150 frames (frames 570–720). Here, the thread instances are mapped to cores in an optimal way for the given layout of worker CPUs and hardware slots. In other words, no other mapping could achieve the user-defined soft real-time constraint for this input data.

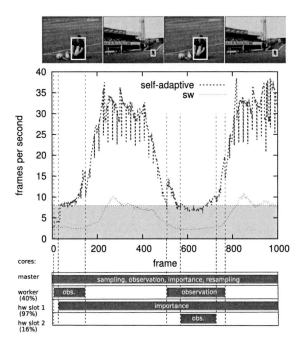

Figure 7.5: Self-adaptation example run considering a soft real-time constraint of eight FPS: Resulting performance in frames/second (upper part) and thread assignment (lower part). Self-adaptation points are represented by vertical dashed lines. The performance target is avoiding the highlighted horizontal bar. The core utilization over time of the worker cores is given in percentage. [8]

7.2.4 Self-adaptation Strategy with Budget

Figure 7.6 shows an example run of our self-adaptive mapping heuristic applied on the hybrid multi-core for the altered soccer video using the ATP_{budget} algorithm (Algorithm 5.4). Again, with an offset of eight frames the self-adaptation technique verifies every twenty frames whether an adaptation is needed based on performance measurements on the master core. The application's performance is measured in frames per second and the desired average performance range is set to eight FPS, where the budget is set to be 33% faster or slower than the defined average performance.

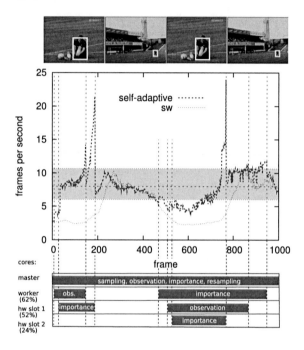

Figure 7.6: Self-adaptation example run: Resulting performance in frames/second (upper part) and thread assignment (lower part). Self-adaptation points are represented by vertical dashed lines. The performance target is highlighted by a horizontal bar. The core utilization over time of the worker cores is given in percentage. [8]

The system starts with all threads running on the master core. Since the resulting performance is lower than the threshold, the first self-adaptation occurs at the eighth frame. At this point, the observation stage takes about 60% of the system's execution time. The worker processor offers the highest speedup for the observation stage and is selected by the algorithm to boost the performance from 2.67 to 4.15 FPS. At the next self-adaptation point (frame 28), the performance still does not meet the performance requirements. Now, the importance stage consumes more than 54% of the system's execution time on the master core. The instantiation of an additional hardware thread promises a total performance of 8.33 FPS, which is close to the desired average performance. Thus, an importance stage thread instance is reconfigured into one of the hardware slots. Due to the reconfiguration overhead, the importance thread becomes available with an offset of eight frames.

After this, the system's performance is inside the requested performance budget for about 120 frames before it exceeds the upper threshold. Now, the system consumes more processing resources than necessary; hence, both worker cores are released again and the application continues with the initial mapping where the entire application runs on the master processor (frame 188). Then the measured performance stays inside the budget for 280 frames, while being entirely executed on the master core. At frames 468, 508, and 528 new cores are added to face the dropping performance, which cannot be completely compensated because the input data complexity overburdens the system for the desired performance constraint. All three worker cores are later deactivated at frames 768, 868, and 948.

This example run stays inside the performance budget for 65.75% of the time, 10.1% of the time the performance exceeds the budget and 24.1% of the time it falls below the budget. The average performance is 7.18 FPS, which is inside the desired performance budget, but below the desired average performance. Compared to the example run in Figure 7.5 the overall worker core utilization reduces from 51% to 46%, where we assume that a worker core is fully utilized in an active state and not utilized in a deactivated state.

7.2.5 Resource Requirements

The resource requirements for the reference design (Figure 7.2) on a ML410 FPGA board are listed in Table 7.4. The reference design consists of a static design including the OS infrastructure, the master CPU, one worker CPU and two reconfigurable hardware cores. Hardware threads have been implemented for the importance and the observation stage, which occupy the same amount of resources for both reconfigurable hardware cores. Note that the resource requirements of the reconfigurable hardware cores only indicate the resources inside the reconfigurable area of the hardware core. The circuit of a hardware thread implementation cannot utilize more than these resources. Here, both hardware threads only utilize a fraction of the available area of the reconfigurable hardware cores.

Table 7.4: Resource requirements for the object tracker case study on a self-adaptive hybrid multi-core (see Figure 7.2) implemented on an ML410 FPGA board.

	LUTs	FFs	BRAMs	DSPs
static design (without HW cores)	12,607	9,876	12	-
observation HW thread	5,917	5,117	11	12
importance HW thread	2,814	1,723	9	15
1st HW core	15,360	15,360	80	40
2nd HW core	16,240	16,240	64	32
ML410	84,352	84,352	376	160

7.3 Thermal Management

This section presents experimental and simulation results on thermal management. The experimental results cover the temperature monitoring, heat generation, sensor self-calibration on modern FPGAs, and, finally, the thermal model of the chip. Furthermore, the simulation results evaluate and compare the developed self-adaptation techniques on hybrid multi-cores using simulation for the non-cooperative and cooperative cases.

7.3.1 Temperature Sensor

The experimental setup for the design space exploration of temperature sensors consists of a Xilinx ML509 FPGA, an external heating device (a hairdryer), and, a workstation connected via UART[23] to the FPGA. The FPGA contains a built-in thermal diode that can be accessed using the Xilinx system monitor. Furthermore, the system monitor can measure the on-chip power supply voltage [158]. The workstation logs the sensor and system monitor readings. The hairdryer is plugged into an USB-controlled power socket, which can be switched on and off by the workstation. The experimental setup is depicted in Figure 7.7.

The sensors are synthesized from a VHDL specification that makes use of device-specific primitives to directly instantiate inverters and enable-logic. In addition to the VHDL description, the ring oscillator's components are placed using placement directives in the user constraints files. This allows the Xilinx embedded development kit (EDK) to pin down individual LUTs and latches to specific slices

---

[23]universal asynchronous receiver transmitter (UART)

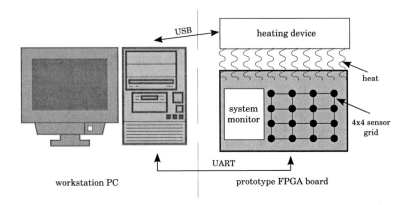

Figure 7.7: Experimental setup: design space exploration of temperature sensors

on the FPGA. The routing is done automatically by the Xilinx EDK tools [160], which employ a randomized algorithm that may produce different results each time a sensor is routed. While this greatly simplifies the task of evaluating a huge number of sensors with different design parameters, the variations in routing have to be accounted for. Hence, each sensor design was evaluated by placing 16 instances of the sensor in a regular 4×4 grid on the FPGA. The resulting values were averaged over the measured sensor performance G (defined in Section 6.1.1) in order to decrease the impact of routing variations.

To evaluate the noise and resolution, two tests were performed for each sensor design. The first test serves to measure the sensor's noise by taking a series of sensor measurements for ten seconds, while keeping the FPGA at a constant temperature. The second test is used to determine the sensor's resolution by heating up the chip an letting it cool down two times over a period of 13 minutes. A wide array of possible ring oscillator designs were examined where the individual layouts differ in the design parameters such as numbers of inverters and numbers of latches. For evaluation the measurement period, the sensor size and the sensor layout were studied.

In order to determine the effect of the measurement period t_m, we measured the noise of sensors of different sizes over varying values for t_m. Figure 7.8 shows the noise as a percentage of the average oscillation count for different measurement periods expressed in clock cycles of a 100 MHz clock. It can be seen that the noise decreases with an increase of t_m from 2^{13} to 2^{16} clock cycles. This is mostly due to quantization noise, which we expect to be at least $\frac{100\%}{t_m} = 0.012\%$ for

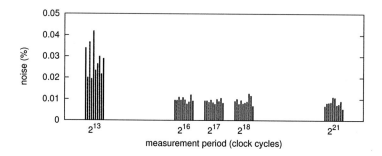

Figure 7.8: Noise (variance in percent) for different sensor sizes and measurement periods. [12]

$t_m = 2^{13}$ clock cycles. A further increase of t_m, however, does not result in a lower noise, as other sources of noise come into play.

Therefore, a measurement period of 2^{16} clock cycles ($655\mu s$) is sufficient for the investigated ML509 FPGA. Short measurement periods have the advantage of limiting the impact of self heating, and the possible advantage of saving reconfigurable resources for the required oscillation counter compared to sensors with longer measurement periods.

In order to examine the influence of the ring oscillator's size we examined several design combinations and evaluated their respective sensor performance. First, we experimented with variations in the number of LUT-based inverter elements. Table 7.5 shows for each sensor design the values of σ_c and σ_v in units of oscillations per measurement. While the noise decreases with the number of inverters, eventually approaching the level of quantization noise, the resolution also decreases, giving rise to a broad optimum in sensor performance at 47 inverters.

As suggested in [170], slice latches that are held in the open state can be built into the oscillator circuit. An advantage of this is that the latches are basically free resources, since on the FPGA each LUT has an associated latch/flip-flop component that can be connected to the LUT's output without the use of additional routing resources. According to Table 7.5 the sensor with 47 elements is the best sensor. Hence, this sensor was used for further design exploration of its layout.

Table 7.5: Average oscillation count, sensor noise, resolution, and performance for different numbers of inverters [12]

# inverters	$\bar{S}$ at 46°C	σ_c	σ_v	G
17	24913	2.4342	17.6854	7.2654
23	18913	1.6990	13.4893	7.9395
31	16263	1.4622	12.1500	8.3092
47	11495	1.0524	8.8124	8.3737
63	8817	0.8218	6.8201	8.2988
79	6956	0.6862	5.4568	7.9526
95	5889	0.5906	4.5791	7.7540
111	5038	0.5200	3.8920	7.4841

Table 7.6: Sensor noise, resolution, and performance for different combinations of LUTs and latches [12]

# latches	# inverters	σ_c	σ_v	G
0	47	1.0524	8.8124	8.3737
16	31	0.9078	7.4713	8.2302
24	23	0.8459	7.9499	9.3977
32	15	0.7668	6.9368	9.0465
38	9	0.7961	6.9750	8.7611
42	5	0.7993	6.7145	8.4001
46	1	0.8270	6.7962	8.2180

Varying subsets of the inverters were replaced with latches to explore the sensor layout. The corresponding results can be found in Table 7.6. The sensor performance G could be improved by up to 13.8% using 23 inverters and 24 latches compared to a sensor layout with 47 inverters and no latches.

While conducting the experiments, the relation between the temperature reported by the Xilinx system monitor and the oscillation frequencies of our sensors differed between the heating up and cooling down periods. Also, there was a variation in on-chip supply voltage as reported by the system monitor. Further investigation revealed that the differences in system monitor temperature with respect to the sensor readings are proportional to the on-chip supply voltage changes.

While we can only speculate about the source of the voltage variations, the dependency of a ring oscillator's frequency on operating voltage is well known and documented. In order to correct the sensor readings for voltage fluctuations, a corrected oscillation count S' was computed from the raw sensor readout S

and the voltage V as reported by the system monitor. The corrected oscillation count can be calculated as follows:

$$S' = S + \xi V \tag{7.1}$$

In order to obtain the coefficient ξ we take advantage of the linear relationship between the temperature and the ring oscillator's frequency and choose ξ in such a manner that this relationship is optimally satisfied.

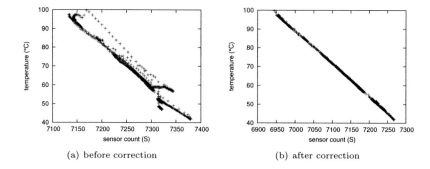

(a) before correction   (b) after correction

Figure 7.9: Example sensor measurements before and after correction. [12]

Figure 7.9(a) shows a set of measurements without correction. Each point represents a raw sensor reading (oscillation count) with the associated system monitor temperature reading. The readings of robust temperature sensors should lie on a line. But as can be seen in Figure 7.9(a), the raw temperature readings seem to lie on curves with different slopes depending on the direction of the temperature changes. Figure 7.9(b) shows the same sample set after the correction is applied. Hence, for robust temperature measurements, the ring oscillator-based temperature sensors should be corrected using Equation 7.1.

7.3.2 Heat Generation

In this section, the temperature measurements for each dedicated heat-generating core are presented and analyzed.

For experimental evaluation, a Xilinx ML509 FPGA board was used. For the LUT- and FF-based heaters, we constrained the area of the heaters to 61×61 slices, which contain $14,884$ LUTs and $14,884$ FFs. This area was used by the LUT oscillator-based heater that contained 1,000 ring oscillators.

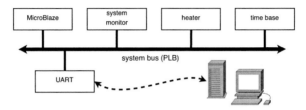

Figure 7.10: Architecture of the experimental setup for heat-generation [10]

The reference architecture is depicted in Figure 7.10. A MicroBlaze processor is connected to the Xilinx system monitor that accesses the sensor readings of the built-in thermal diode on the FPGA. Furthermore, the architecture contains a time base and the heat-generating core under examination. The heater is enabled/disabled by a timer-driven program that runs on the MicroBlaze, which also reads the temperature values. The temperatures readings are forwarded to a workstation using a UART interface.

In all experiments, the system first waits for 700 seconds until the temperature of the FPGA is stable, before the heater is enabled for 700 seconds. Next, the system disables the heater for 700 seconds to see the temperature decrease for the cooling phase. Finally, the system enables the heater again for another 700 seconds to confirm the repeatability of the experiment. Figure 7.11 shows the temperature measurements for the heaters that have been introduced in Section 6.2.

For the first heater that is based on LUT pipelines, 14 pipelines were implemented with $1,000$ pipeline stages. The corresponding measurement is depicted in Figure 7.11(a). A 100 MHz clock signal is the input for all 14 pipelines. It can be seen that the FPGA temperature increases about $3°C$ in 700 seconds.

As can be seen in Figure 7.11(b), a LUT oscillator heater with 1,000 oscillators already heats up the FPGA to $195°C$ according to the built-in thermal diode. The experiments were restricted to 1,000 ring oscillators in order to prevent the destruction of the FPGA. To be able to compare the diverse LUT- and FF-based heaters with each other, the same amount of slices was used for each LUT- and FF-based heater. By replacing the 14 pipelines with 1,000 ring oscillators (each implemented with a single LUT) instead, the FPGA heats up to about 195°C, which represents a temperature increase of about 135°C in 700 seconds.

Both cores toggle 1,000 signals. However, while the LUT-based ring-oscillators toggle their output signals at the highest possible frequency, the LUT pipelines toggle their signal at 100 MHz only. Because of the poor heat output, further experiments were omitted for the LUT pipelines.

Figure 7.11(c) shows the measurements for the SRLs. The SRL pipeline heater has shown superior results to the LUT pipeline heater when considering that it utilizes only a fraction of the LUTs, because SRLs can only be mapped to specific slices of the Virtex-5 FPGA. Therefore, only 41 SRL pipelines with 100 pipeline stages could be mapped into the constrained area. Thus, the heater only consumed 4,100 LUTs instead of 14,000 LUTs that are used for the LUT pipelines.

For the heaters depicted in Figures 7.11(c)-7.11(h) the influence of the clocking frequency to the heat generation was quantified for the ML509 FPGA board. The lowest clock frequency for all experiments was 100 MHz. Then, the clock frequency was increased in steps by 100 MHz. For the heaters that employ SRLs the maximum operational frequency is 300 MHz. The FF and LUT-FF pipelines could be clocked up to 400 MHz. Finally, the maximum operational frequency is 500 MHz for the BRAM heater and 550 MHZ for the DSP heater, respectively.

For the FF heater, again 14 pipelines were implemented with 1,000 stages each. Similar to the experiments for the SRL pipelines, see Figure 7.11(c), an increase in the clocking frequency also increases the heat on the FPGA. The FF heater could increase the FPGA temperature between 5°C and 22°C depending on the clock rate, as can be seen in Figure 7.11(d).

Hybrid heaters that use both, LUTs and FFs, clearly outperform the heaters that either use LUTs or FFs when the resources implement clocked pipelines. This can be seen in Figure 7.11(e) for the LUT-FF pipelines. The LUT-FF heater contains 14 pipelines with 1,000 stages where each stage contains a LUT and a FF. The result is similar for the SRL-FF heat core, which is depicted in Figure 7.11(f). Note that again the SRL-FF heat core uses 41 pipelines with 100 stages and, thus, consumes significantly less resources than the LUT-FF heater.

Figure 7.11(g) shows the measurements from the BRAM heater. Here, we were able to clock the heater up to 500 MHz. The BRAM heater contains 130 BRAMs and can heat up the FPGA from 5°C up to 25°C depending on the clocking frequency. Finally, the DSP heater contains a single pipeline of 38 DSPs. This

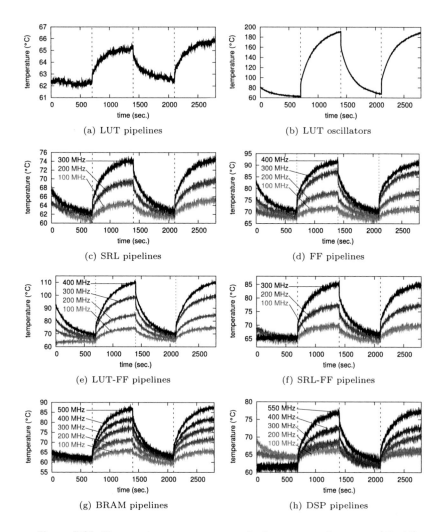

Figure 7.11: Temperature measurements for heat-generating cores. [10, 62]

time we were able to clock the heater up to 550 MHz; the results can be seen in Figure 7.11(h).

For the DSP blocks, the temperature increase in 700 seconds again depends on the clocking frequency. Here, the increase was 2°C for 100 MHz and 14°C for 550 MHz. However, this pipeline additionally introduced a high number of LUTs and FFs and it might be the case that most of the temperature is generated by these resources. This seems plausible if the measurement results for the heaters that are purely based on LUTs and FFs are taken into account.

Table 7.7 lists the resource utilization of the different heat-generating cores. Note that each heater includes a bus attachment to the processor local bus (PLB), which utilizes a few LUTs and FFs. This can be seen, i.e., for the LUT pipelines that consume 248 additional LUTs and 376 FFs. The LUT oscillators required additional logic to ensure that the 1-level ring oscillators are connected to the system, so that they were not trimmed by the place and route tools. Finally, the DSP pipelines use additional LUTs and FFs for auxiliary logic.

Table 7.7: Resource utilization for the heaters [10]

heater	LUTs	FFs	BRAMs	DSPs
LUT pipelines	14,248	376	-	-
LUT oscillators	14,608	375	-	-
SRL pipelines	4,364	376	-	-
FF pipelines	233	14,376	-	-
LUT-FF pipelines	14,276	14,376	-	-
SRL-FF pipelines	4,408	4,476	-	-
BRAM pipelines	223	311	130	-
DSP pipelines	2,252	4,310	-	38
Virtex-5 LX110t	69,120	69,120	256	64

The temperature increases for all heaters over 700 seconds are summarized in Table 7.8. The two highest temperature increases are generated by the LUT oscillators and the LUT-FF heater at 400 MHz. In general, the advantage of heaters, which are based on LUTs or FFs is their high flexibility. They can have arbitrary size and can be placed almost anywhere on the chip. In contrast to LUTs getting their input from a system clock, SRLs show better results for heat creation if resource utilization is taken into account. However, the amount of LUTs that can implement SRLs is limited on the Virtex-5. Hence, for the same rectangular area constraint, less heat can be generated for SRLs. DSPs blocks show poor results for heat creation, since a significant amount of generated heat is likely to come from the surrounding logic. In contrast, the BRAMs can heat up the FPGA on their own.

Table 7.8: Temperature increase of the heaters over 700 seconds [10]

heater	temperature rise			
LUT	+3°C			
LUT osc.	**+134°C**			

heater	temperature rise with different frequencies [MHz]				
	100	**200**	**300**	**400**	**500/550**
SRL	+4°C	+7°C	+12°C	-	-
FF	+5°C	+11°C	+18°C	+22°C	-
LUT-FF	+10°C	+15°C	+31°C	**+41°C**	-
SRL-FF	+6°C	+12°C	+20°C	-	-
BRAM	+5°C	+10°C	+14°C	+20°C	+25°C
DSP	+2°C	+4°C	+7°C	+11°C	+14°C

7.3.3 Sensor Self-calibration

For experimental measurements of spatial temperature differences, a Xilinx ML605 FPGA board was used that contains a built-in thermal diode, which can be accessed inside the FPGA using the dedicated system monitor hard macro. The FPGA contains 160×239 slices, which were partitioned into a regular grid of 10×15 tiles. Each tile contains a temperature sensor at the center. For experimental evaluation, this setup proved to be a good compromise between grid resolution and FPGA resource usage. On this FPGA, there is a central region which is not reconfigurable so that we could not place temperature sensors for six tiles in this region.

Figure 6.1 shows the employed sensor grid. Twelve heat-generating circuits were implemented and constrained to disjunct areas that can be seen in Figure 6.2. This layout was chosen because, on the one hand, it enables us to heat up the left, right, upper and lower sides of the FPGA independently and, on the other hand, the size and shape of the heater regions resemble that of possible hardware threads that may run on the FPGA. One regional heater was implemented using 10,000 flip-flops that toggle at a frequency of 100 MHz.

In order to calibrate the sensors, the system must first be in thermal equilibrium, which we assume is reached when the thermal diode does not vary more than

0.3°C in a time interval of 20 seconds. Then, the system makes a temperature measurement for each sensor and stores the number of oscillations for a fixed time interval of 2^{17} clock cycles as well as the measured temperature of the thermal diode. In a second step, the FPGA activates all heat-generating circuits and again waits for thermal equilibrium. At this point, the system makes a second temperature measurement for each sensor and computes a linear function mapping sensor readings to temperature. The sensor self-calibration takes about 3 to 4 minutes in our experiments.

Figure 7.12 depicts an example temperature distribution that was measured after self-calibration. In all the experiments of Sections 7.3.3-7.3.4, the fan of the FPGA's cooling element is active.

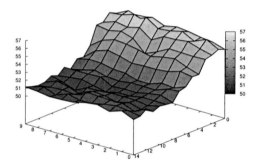

Figure 7.12: Measuring temperature distribution (in °C) on an Virtex-6 FPGA where the top five heat-generating cores are activated. The x- and y-axes represent the area of the chip, which is partitioned into 10 × 15 tiles, and the z-axis represents the tile temperatures. [6]

7.3.4 Thermal Model

After sensor calibration, the temperature model parameters are learned. Therefore, the system creates a 12-minute test scenario, see Table 7.9, where the system creates various spatial temperature differences over time. In a first step, the system activates all heaters for two minutes, before it deactivates them again. In a second step, it activates the regional heaters on each side of the FPGA for one minute, before it deactivates them again for one minute. In this scenario, the system measured spatial temperature differences up to 6.5°C (see Figure 7.12) and a temperature difference in time up to 9°C. The temperature reading of a specific sensor (marked in Figure 6.1) can be seen in Figure 7.13. The system performs measurements on the entire sensor grid each second and stores all temperature readings in the main memory.

Table 7.9: 12-minute learning scenario [6]

time (min.)	description
1-3	all regional heaters are activated
4-5	top five regional heaters are activated
6-7	bottom five regional heaters are activated
8-9	five regional heaters to the right are activated
10-11	five regional heaters to the left are activated
other	all regional heaters are deactivated

Then, randomized hill climbing (cf. Section 6.3.1) is applied for the measurement data. Table 7.10 shows the initial set P_{init} of the free parameters for the learning algorithm that was defined manually. The parameters $C(i)$, $i \in L_0$ and T_s have been set to 0.001 and to 25, respectively. Then, the other parameter have been adapted by hand until the simulated temperatures reached a point where the prediction error was less than $10°C$.

Note that the individual parameter values do not necessarily represent the physical characteristics. Equations 6.5-6.7 in Section 6.3 define the temperature change of a tile over time, where the temperature of a considered tile is influenced by a possible heat source of a thread, the temperature of neighboring tiles and the heat flow to the heat sink. The considered tile is connected to its neighboring tiles, the heat source and the heat sink and receives the corresponding temperature differences multiplied with a specific $\frac{1}{RC}$ coefficient. The learning algorithm can only learn the product of the specific resistance R and the tile's capacitance C. This means that there can be multiple different parameter sets which result in the same thermal simulation.

For instance, when each capacitance C of the model is multiplied by an arbitrary factor $k \neq 0$ and all resistances R is divided by k, the thermal models of the initial parameter set and the adapted parameter set are identical. Hence, the model parameters do not necessarily represent the physical thermal resistances or capacitances of the system.

Table 7.11 defines the improvement of the average root mean square error (RMSE) between measurement and simulation data while the parameters are learned for both stages.

The temperatures of the tiles in layer 0 are initialized with the first measurement. For this scenario the number of heat sources $I_{source}(i)$ is limited to the cases

Table 7.10: Initial and learned temperature model parameter. [6]

param.	P_{init}	$P_{learned}$	param.	P_{init}	$P_{learned}$
$R_v(i)$	100	101.07	$C(i), i \in L_0$	0.001	0.00099
$R_s(i)$	33.333	35.164	$C(i), i \in L_1$	1.5	1.51013
$R_{x,0}$	150	151.87	$I_{on}(i)$	0.25	0.22636
$R_{y,0}$	150	148.47	$I_{off}(i)$	0.15	0.17421
$R_{x,1}$	0.0667	0.0682	T_s	25	25.2999
$R_{y,1}$	0.0667	0.0681			

where the regional heater that covers tile i is activated, $I_{on}(i)$, and where the regional heater is deactivated, $I_{off}(i)$.

Table 7.11: Learning progress of the randomized hill climbing algorithm [6]

stage	RMSE in (°C)
initial	3.256703
stage 1	0.773082
stage 2	0.719692

Here, the learning algorithm is executed on a MicroBlaze processor clocked at 100 MHz. Learning the thermal model parameters at run-time takes between 50 and 60 minutes depending on the input data in our experiments. The learned (averaged) temperature model parameters $P_{learned}$ are listed in Table 7.10. It can be seen that the learned parameter set is similar to the initial manually-defined parameter set. However, the prediction results show that the thermal model has been considerably improved due to learning.

Figure 7.13 compares the temperature measurements with the temperature predictions according to the learned temperature model for a selected sensor. It can be seen that the temperature predictions closely match the measured temperatures. The average prediction error for all sensors is 0.72°C for this 12-minute scenario. Predicting the entire scenario on a MicroBlaze processor clocked at 100 MHz at run-time takes 99.5 seconds. A time resolution of $\Delta t = 0.02$ seconds was used for a prediction step. The prediction time can be reduced if the time resolution is reduced, but this has negative effects on the prediction accuracy.

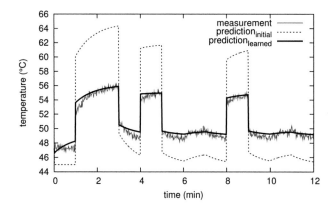

Figure 7.13: Temperature difference between measured and predicted temperature over the entire scenario sample set (defined in Table 7.9) for a selected sensor. The sensor position is marked in Figure 6.1. [6]

7.3.5 Self-adaptation Strategies

This section presents simulation results for the non-cooperative and cooperative self-adaptation strategies, which have been presented in Section 6.4. We apply simulation because this allows for an efficient evaluation of various multi-core layouts, thread sets, and workload scenarios without the burden of implementing these designs on an FPGA. Furthermore, we can simulate core layouts that can currently not be implemented using ReconOS, such as multi-cores with more than 15 cores.

In Section 7.3.4 the experimental results for learning the thermal model of a specific Virtex-6 FPGA have been presented. The thermal model was learned for an FPGA with an active cooling element (fan). The underlying thermal model, which is used for the simulations, has been learned on the same FPGA but without the active cooling element. Active cooling limits the heat flow between the tiles of the FPGA and reduces the maximum temperature. Hence, the thermal model for an FPGA without active cooling is more challenging and, therefore, more interesting for studying thermal management techniques. The prediction/simulation accuracies of the thermal models (with and without active cooling) do not differ significantly. Thus, the results for the thermal model of an FPGA without active cooling have not been presented in this thesis. The FPGA was partitioned into a regular grid of 10×15 tiles. For the thermal simulations two different multi-core layouts have been studied, which can be seen in Figure 7.14.

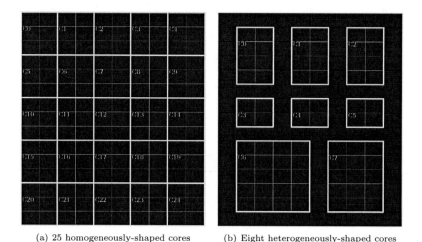

(a) 25 homogeneously-shaped cores   (b) Eight heterogeneously-shaped cores

Figure 7.14: Simulated multi-core layouts

The first multi-core layout, depicted in Figure 7.14(a), consists of 25 homogeneously-shaped cores. Each core covers six tiles of the FPGA. The second layout is depicted in Figure 7.14(b) and consists of eight cores, which have heterogeneous shapes. The individual cores can stretch across either four, eight, fifteen, or twenty tiles. There are three cores with four tiles and three cores with eight tiles. The heterogeneous shapes of the eight-core layout model the possibility that the hybrid multi-core consists of heterogeneous hardware and software cores, which require different amounts of resources. For instance, the cores C_3, C_4 and C_5 might represent processors whereas the other cores might represent hardware cores, where two of the hardware cores cover more tiles than the others. In contrast to the 25-core, not all tiles are covered by cores in the eight-core layout. These tiles might contain additional peripherals, system buses or no logic. The heat generation of the logic that is not part of the cores is omitted in all simulations.

Two scenarios have been studied. In the first scenario the sum of all heat source differences $\Delta\phi(T_i, C_j, \Lambda_l)$ of a thread $T_i \in T$ is the same for all cores:

$$\forall C_j, C_k \in C : \sum_{\Lambda_{l_1} \in \lambda(C_j)} \Delta\phi(T_i, C_j, \Lambda_{l_1}) = \sum_{\Lambda_{l_2} \in \lambda(C_k)} \Delta\phi(T_i, C_k, \Lambda_{l_2}) \qquad (7.2)$$

This condition models that each thread consumes the same power on each core.

In the second scenario, the cores $C_1, \ldots, C_{|C|}$ have individual heat coefficients $\theta_1, \ldots, \theta_{|C|}$. This models a scenario where the cores have different thermal characteristics. For instance, the cores can have different clock frequencies. In this scenario the heat coefficients are the same for all threads in T. Hence, for all threads $T_i \in T$ the following condition holds.

$$\forall C_j, C_k \in C : \sum_{\Lambda_{l_1} \in \lambda(C_j)} \frac{\Delta\phi(T_i, C_j, \Lambda_{l_1})}{\theta_j} = \sum_{\Lambda_{l_2} \in \lambda(C_k)} \frac{\Delta\phi(T_i, C_k, \Lambda_{l_2})}{\theta_k} \qquad (7.3)$$

In the performed simulations the heat coefficients were selected randomly from the interval $[0.85, 1.35]$ for the 25-core and from the interval $[0.85, 5.85]$ for the eight-core using a uniform distribution. For all thermal simulations the execution times of each thread is independend of its mapping to a specific core. For a more accurate simulation the speedup values $S_{i,j}$ of the individual thread-to-core mappings have to be considered. However, a joint performance and thermal management for hybrid multi-cores is out of scope of this thesis. Furthermore, the overhead of thread migration in time and heat and further system components such as the system buses, memory controllers, and peripherals are neglected for all simulations.

For both scenarios, 100 simulations have been performed for different workloads and thread sets. The workload is given by a timetable, which defines at which point in time which thread gets new workload. The workload is given in amount of seconds that a thread has to be executed. For the 25-core the duration between new workloads is selected randomly between 4 and 12 seconds and for the eight-core it is selected between 20 and 50 seconds. The duration is not constant for a simulation, but the duration is randomly selected for each new occurrence of workload. However, on average the number of time steps with new workload are approximately the same for each simulation. One to three threads receive new workload at a time step. The workload for both scenarios is randomly selected from the interval $[10, 100]$. The thread, which receives the new workload, is selected randomly among the thread set.

The thread set is given by the heat sources for each (thread,core) combination as defined in Section 6.4. The heat sources do not represent exact physical variables but define an input for our thermal model, which allows accurate temperature simulations. The heat sources are selected randomly inside intervals $[0.35, 0.7]$ for the 25-core and $[0.45, 0.9]$ for the eight-core using a uniform distribution. The default heat sources of the underlying thermal model, where all cores idle, is approximately 0.3 for each tile. These default heat source values have been learned on an actual Virtex-6 FPGA. Note that the default heat source values are higher than the default heat sources I_{off} in Section 7.3.4 because the active cooling has been disabled. Passive cooling results in higher temperatures and, therefore, higher heat sources compared to active cooling with a fan. The maximum tile temperature in the simulations has been $112.5°C$, which is far below the temperatures that could be measured on a Virtex-5 FPGA using dedicated heat-generating cores, see Section 7.3.2. Hence, we believe that the selected intervals for heat sources are justifiable.

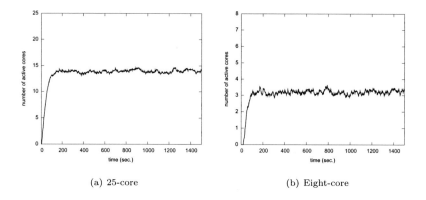

(a) 25-core   (b) Eight-core

Figure 7.15: Average workload for scenarios 1 and 2

A simulation series with 100 different workloads and thread sets have been generated and evaluated using the different strategies. For each simulation series, heat sources of 100 threads have been generated. The simulation duration has been set to 1500 seconds. All presented results for the entire simulation series are averaged. In our simulations, it could be observed that the temperature intervals of the chip stabilize after a duration of about 1,000 seconds. At the beginning the chip temperature always starts at about $90°C$ for each tile and then increases due to the workload. The number of active cores of the individual simulation series can be seen for both core layouts in Figure 7.15. The average number of active cores is about 14 after 200 seconds for the 25-core and approximately three for the eight-core. Hence, on average there are usually free cores that can be used for non-cooperative and cooperative thread-to-core re-mappings.

The non-cooperative and cooperative self-adaptation strategies are listed in Table 7.12. Eleven cooperative strategies combine heuristics for the initial thread-to-core assignment, heuristics for the detection of the hottest core, and heuristics for the dynamic thread migration between cores. All heuristics have been defined in Section 6.4. Because the cooperative strategies use the same heuristics for thread assignments as the non-cooperative strategies, the benefits of dynamic thread migration can be evaluated for the considered core layouts and scenarios.

Table 7.12: Non-cooperative and cooperative self-adaptation strategies

heuristic	non-cooperative		cooperative	
	assignment	hottest core	assignment	migration
h_1	$h_{n,\text{naive}}$	-	$h_{n,\text{naive}}$	$h_{p,\text{rotate}}$
h_2	$h_{n,\text{random}}$	-	$h_{n,\text{random}}$	$h_{p,\text{random}}$
h_3	$h_{n,\alpha,\text{max}}$	$h_{\sigma,\text{max}}$	$h_{n,\alpha,\text{max}}$	$h_{p,\alpha,\text{max}}$
h_4	$h_{n,\alpha,\text{sum}}$	$h_{\sigma,\text{sum}}$	$h_{n,\alpha,\text{sum}}$	$h_{p,\alpha,\text{sum}}$
h_5	$h_{n,\alpha,\text{avg}}$	$h_{\sigma,\text{avg}}$	$h_{n,\alpha,\text{avg}}$	$h_{p,\alpha,\text{avg}}$
h_6	$h_{n,\beta,\text{max}}$	$h_{\sigma,\text{max}}$	$h_{n,\beta,\text{max}}$	$h_{p,\beta,\text{max}}$
h_7	$h_{n,\beta,\text{sum}}$	$h_{\sigma,\text{sum}}$	$h_{n,\beta,\text{sum}}$	$h_{p,\beta,\text{sum}}$
h_8	$h_{n,\beta,\text{avg}}$	$h_{\sigma,\text{avg}}$	$h_{n,\beta,\text{avg}}$	$h_{p,\beta,\text{avg}}$
h_9	$h_{n,\gamma,\text{max}}$	$h_{\sigma,\text{max}}$	$h_{n,\gamma,\text{max}}$	$h_{p,\gamma,\text{max}}$
h_{10}	$h_{n,\gamma,\text{sum}}$	$h_{\sigma,\text{sum}}$	$h_{n,\gamma,\text{sum}}$	$h_{p,\gamma,\text{sum}}$
h_{11}	$h_{n,\gamma,\text{avg}}$	$h_{\sigma,\text{avg}}$	$h_{n,\gamma,\text{avg}}$	$h_{p,\gamma,\text{avg}}$

The thermal imbalance of the eight-core over time is depicted for the cooperative heuristics h_1, h_{10} and h_{11} in Figure 7.16. The graphs show the interval between the hottest and the coldest tile temperature for each second, which defines the thermal imbalance on the chip over time. Furthermore, the average tile temperature over time is drawn using a single line. In this figure, the thermal imbalance on the chip over time is roughly the same for the heuristic h_1 and h_{10}. However, the thermal imbalance between the maximum and minimum tile temperatures vary significantly. As previously described the thermal imbalance does stabilize after approximately 1,000 seconds. Hence, in the following only the average thermal imbalance for the time interval $[1300, 1500]$ is depicted to allow for an intuitive comparison between all non-cooperative and cooperative heuristics (see Figures 7.17-7.20).

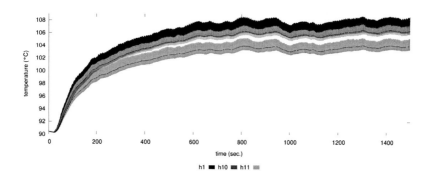

Figure 7.16: Thermal imbalance of the eight-core over time for scenario 2 using a subset of the cooperative self-adaptation strategies

Homogeneously-shaped 25-core

Figure 7.17(a) shows the simulation results for the homogeneous 25-core, where each thread T_i creates a specific additional heat source $\Delta\phi()$, which is independent of the core C_k that executes T_i. Hence, the heat source values $\phi()$, which also include the static heat source of the idle thread, only differ marginally. According to our experimental results on a Virtex-6 FPGA, not all tiles have the same temperature when the FPGA-based multi-core does not execute any application. This imbalance can be caused by process variation on transistor level. Thus, our thermal model learns the individual idle heat sources for each tile individually. As a result, the heuristics that either fully (h_6-h_8) or partly (h_9-h_{11}) base their decision on the heat source values, can differ in their decisions.

The naive mapping strategy provides the worst results and creates a thermal imbalance of $3.2°$C on the chip, which is defined as the maximum tile temperature minus the minimum tile temperature of the chip. This is because the naive non-cooperative mapping strategy clusters the threads together. Inside the cluster of active cores the heat cannot distribute to colder cores and, therefore, the maximum tile temperature is higher than for any other cluster. Furthermore, there can be another cluster of inactive cores, where some cores are not adjacent to any active heater, even when 14 out of 25 cores are active. Applying the random non-cooperative strategy h_2 already reduces the thermal balance to $1.9°$C.

The strategies h_6-h_8 map according to the heat source values, where the thermal imbalance on chip is close to the one achieved by random mapping. Here, h_6

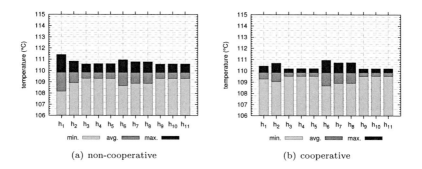

Figure 7.17: 25-core: Thermal simulation results for scenario 1

results in a thermal imbalance of 2.3°C and h_7-h_8 of 1.9°C. Note that for the homogeneously-shaped 25-core the heuristics applying the sum or the avg criterion evaluate the cores in the same manner. Using the homogeneous 25-core layout, all cores have the same amount of tiles. When the cores are now ordered by their temperature readings, heat sources, or the product of both, the ordering is the same for the average criterion and the sum criterion. The only difference is that the avg criterion divides its core evaluation values by the number of tiles of the specific core. When this number is constant, the ordering is the same for both criteria. Hence, both criteria select the same core, unless there are multiple candidates with the best evaluation value. Taking into account the current temperature distribution of the chip limits the thermal imbalance to 1.3°C with heuristics h_3-h_5 and h_9-h_{11}.

By introducing cooperative thread migration, the thermal imbalance can be lowered to 0.7°C when the heuristics base their mapping decisions fully or partly on the temperature, see Figure 7.17(b). The number of cooperative thread migrations is listed for both core layouts and both scenarios in Table 7.13. The heuristics that only partly base their mapping decision on the temperature (h_9-h_{11}) perform fewer thread migrations but achieve the same thermal balance. The heuristic with the most thread migrations is the rotating heuristic h_1, which rotates each thread to a new core in each second. About 27,700 thread migrations are performed. However, the thermal imbalance of the initial naive mapping can only be reduced to 1.1°C.

The random cooperative strategy h_2 performs about 2,280 random migrations and reduces the thermal imbalance of the initial random mapping only by 0.3°C. It can be seen that for this core layout and this scenario the temperature-driven

Table 7.13: Average number of cooperative migrations in 1500 seconds

heuristic	eight-core		25-core	
	scenario 1	scenario 2	scenario 1	scenario 2
h_1	27,662.19	27,733.17	6,896.00	6,834.18
h_2	2,273.45	2,279.10	1,903.77	1,890.66
h_3	1,658.05	1,660.48	1,482.41	1,496.28
h_4	1,655.11	1,656.53	1,021.79	989.59
h_5	1,655.22	1,656.50	1,498.87	1,512.19
h_6	17.42	102.16	76.35	73.12
h_7	63.83	110.43	23.10	50.70
h_8	66.66	109.56	78.51	74.29
h_9	1,514.83	1,512.60	1,265.70	1,244.72
h_{10}	1,563.37	1,559.05	1,269.34	1,004.94
h_{11}	1,563.50	1,559.29	1,275.73	1,294.88

mapping strategies outperform the other strategies. The average temperature of the chip, however, is almost equal for all heuristics in the first scenario. This is because the simulations presented in Figure 7.17 are performed on a homogeneous multi-core, where each core has the same size and each core generates the same heat for a specific thread.

In contrast to the previous scenario, the second scenario considers a heterogeneous multi-core, where the heat sources of a thread depend on the core that executes the thread. The simulation results for the non-cooperative mapping strategies are depicted in Figure 7.18(a). Again, the naive mapping strategy h_1 delivers the highest thermal imbalance on chip. The random heuristic h_2 results in a lower thermal imbalance than the heat source-driven heuristics h_6-h_8. However, the heat source-driven heuristics generally reduce the thermal profile of the chip. When considering heterogeneous cores, the heat source-driven heuristics h_6-h_8 significantly reduce the average tile temperature of the chip by about $1.5°C$ compared to the heuristics h_2-h_5 and h_9-h_{11}. This is because the heuristics h_6-h_8 assign the threads to the most efficient cores, such that the overall temperature can be considerably reduced. The temperature-driven heuristics h_3-h_5 achieve a lower thermal imbalance of $1.4°C$ compared to $2.3°C$ achieved by h_6-h_8. The final heuristics h_9-h_{11} appear to be a compromise between the temperature-driven and the heat source-driven non-cooperative mapping strategies.

Thread migrations can lower the thermal imbalance on chip as can be seen in Figure 7.18(b). The rotating heuristic h_1 can limit the thermal imbalance to

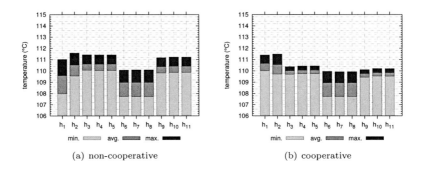

Figure 7.18: 25-core: Thermal simulation results for scenario 2

1.4°C at the price of about 27,800 thread migrations. The heuristics that take the tile temperatures into account show the best results in order to reduce the thermal imbalance at the price of 1,500-1,660 thread migrations. Again, the additional consideration of the heat sources saves about 100 thread migrations. The heat source-driven heuristics rarely migrate. Compared to the first scenario, they migrate more often. The heat source-driven heuristics only remap in case the initial mapping strategy could not map the new thread to its best core, because another thread already runs on this core. In this case, the heuristic might decide to remap, whenever this seems beneficial. Because of the rare remappings, the heuristics h_6-h_8 do not decrease the thermal imbalance.

Heterogeneously-shaped eight-core

Next, the simulation results for the heterogeneously-shaped eight-core are presented. The heterogeneous shapes of the cores model a hybrid multi-core, which contains processors and reconfigurable hardware cores. The reconfigurable hardware cores can have different sizes in order to provide more logic resources to a hardware implementation of a thread or to decrease the thermal footprint of a hardware thread. In our simulations the sum of additional heat sources of the threads is the same in the first scenario. In the second scenario each core C_j has a heat coefficient θ_j, which is multiplied by the heat source values of all threads that are mapped on C_j.

The non-cooperative mapping results are presented in Figure 7.19(a). The average tile temperature is similar for all heuristics. Applying the naive mapping h_1 results in the highest thermal imbalance of 3.0°C, whereas applying the heuristic

h_7 results in a thermal imbalance of 2.7°C. For all other heuristics the thermal imbalance is about 2.3°C. Due to the assumptions of the first scenario, the power of a thread does not depend on its mapping. Hence, the average temperature is the same for all heuristics.

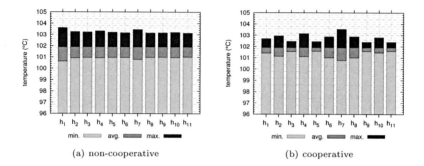

(a) non-cooperative

(b) cooperative

Figure 7.19: Eight-core: Thermal simulation results for scenario 1

Allowing cooperative thread migrations each second reduces the thermal imbalance of the chip, see Figure 7.19(b). This time, the cooperative rotating heuristic h_1 reduces the thermal imbalance from 3.0°C to 1.3°C at the cost of 6,896 thread migrations, see Table 7.13. The random cooperative heuristic achieves a thermal imbalance of 1.8°C by using about 1,900 thread migrations. In contrast to that, the heuristics h_3, h_5, which take the temperature readings into account, reduce the imbalance to only 0.9°C using about 1,500 thread migrations. The heuristics h_9 and h_{11} achieve similar thermal results with less thread migrations by additionally taking into account the heat sources. It can be seen that the max and avg criteria, e.g., h_3 and h_5, outperform the sum criterion, e.g., h_4.

Figure 7.20 depicts the results of the non-cooperative strategies for the second scenario. Once more the naive mapping h_1 creates the highest thermal imbalance of about 5.2°C. The heuristics h_4, h_7 and h_{10}, which use the sum criterion to evaluate the mappings, significantly improve the thermal profile of the chip by lowering the overall temperature of all tiles. The heat source-driven heuristic h_7 provides the best results. All heuristics that apply the max and avg criteria provide better results compared with naive mapping and random mapping, on the one side, but are clearly outperformed by the heuristics that apply the sum criterion, on the other side. This indicates that the average tile temperature or the maximum tile temperature of a core should not be used to drive the thermal re-mapping of a heterogeneously-shaped multi-core in the second scenario, because they do not take the core sizes into account. In this scenario, the heat sources of a thread are lower on large cores than on small cores. However, because the sum of the heat source differences can be the same for different core sizes, this does not imply that large cores are necessarily more suited for lowering the overall chip temperature. In contrast to the `max` and `avg` criteria, the `sum` criterion respects this knowledge.

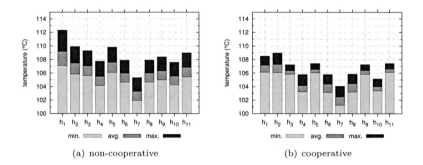

(a) non-cooperative   (b) cooperative

Figure 7.20: Eight-core: Thermal simulation results for scenario 2

Allowing cooperative thread migrations each second reduces the thermal imbalance of the chip, see Figure 7.19(b). This time, the heuristics that are based partly or fully on the temperature and use the `max` and `avg` criteria, i.e., h_3,h_5,h_9, and h_{11}, lower the thermal imbalance to approximately 1.4°C. These heuristics even outperform the rotating heuristic h_1 by using only a fraction of the thread migrations, approximately 1500 compared to approximately 6850. The `sum` criteria applied in heuristics h_4, h_7 and h_{10} further lower the temperature profile of the chip compared to their non-cooperative counterparts. However, by using only about 1,000 thread migrations for h_4 and h_{10} and by using only about 50 thread migrations for h_7, they cannot balance the chip temperature as good as the heuristics that apply the `max` and `avg` criteria. Again, the heat source-driven cooperative heuristics only perform a few thread migrations. Hence, the `max` criterion or the `avg` criterion should be applied, when on-chip temperature distribution has to be balanced. In contrast, when the overall thermal profile of the chip must be lowered, the `sum` criterion should be used.

In summary, the cooperative heuristics h_3,h_5,h_9, and h_{11}, which are fully or partly based on the temperature readings, provide the best results for limiting the thermal imbalance on chip. The heuristics h_9 and h_{11}, which additionally take the heat sources into account, lower the number of performed thread migrations. Hence, allowing cooperative thread migrations supports balancing the on-chip temperature distribution. Furthermore, all heuristics that consider the different sizes of the cores decrease the overall chip temperature in the second scenario for a heterogeneously-shaped multi-core. The cooperative heat source-driven heuristic h_7 achieves the best results for lowering the overall temperature profile of the chip by performing a rather small number of thread migrations in the second scenario. The numbers of thread migrations can be reduced by at least an order of magnitude compared to the heuristics, which are not fully driven by the heat sources.

7.4 Chapter Conclusion

This chapter presented an experimental evaluation of the proposed concepts, models, and algorithms of self-adaptive hybrid multi-cores on FPGA-based prototyping platforms. Xilinx Virtex FPGAs were used for implementation of the self-adaptive hybrid multi-cores because they support dynamic partial reconfiguration. Compared to other Xilinx FPGA families, the Virtex FPGA family provides a huge amount of programmable logic, which allows for large designs.

For performance management, a multi-threaded video object tracker was analyzed as a case study for streaming applications. The threads of the application were mapped to a self-adaptive hybrid multi-core that consisted of two PowerPC processors and two reconfigurable hardware cores. The hybrid multi-core was implemented using the common programming model and execution environment ReconOS. The performance model of the application was known a priory. It could be shown that by monitoring the performance of the threads at run-time, the hybrid multi-core was able to adapt its thread mapping in order to control the application's performance. The tracking performance (in time) strongly depends on the incoming video frames. Two different self-adaptation techniques were successfully employed. The first strategy ATP_{bound} provides a user-defined lower performance bound and the strategy ATP_{budget} meets a user-defined performance interval for the majority of the time while both strategies minimize the number of active cores at the same time (in order to save power or to free the cores for other applications).

For thermal management, experimental results show that a ring oscillator-based temperature sensor grid can be (self) calibrated without the need of external devices such as a temperature-controlled oven or an infrared camera. To obtain good sensors, an extensive design space exploration of ring oscillator-based sensors was presented. Experiments show that regional heat-generating cores can be used to globally heat up the chip and to calibrate the sensors of a 10×15 sensor grid against a built-in thermal diode in four minutes on a Xilinx Virtex-6 FPGA. It could be shown that spatial temperature differences can be measured on today's FPGAs. Spatial temperature differences among the entire chip of up to 6.5°C could be generated using toggling flip-flops. A further investigation on regional heat generating cores revealed that the FPGA temperature can be increased by up to 134°C on Virtex-5 FPGA by only utilizing about 20% of the look-up-tables. The thermal model of a Virtex-6 FPGA was learned using randomized hill climbing with an average root mean square error of 0.72°C. The execution time of the learning phase was about 50 to 60 minutes on an embedded MicroBlaze processor clocked at 100 MHz. Using the thermal model, the system can predict the temperature distributions of different activation patterns of the regional heaters.

In a final step, the learned thermal model was used inside a temperature simulator for self-adaptive hybrid multi-cores to evaluate eleven different non-cooperative and cooperative self-adaptation strategies. Two different core layouts, a homogeneously-shaped 25-core and a heterogeneously-shaped eight-core were analyzed. For both core layouts, two different scenarios were investigated. In the first scenario, each thread employed the same heat sources on all cores. In the second scenario, the cores had individual heat coefficients, which scaled the heat sources of the threads. Hence, some cores generally generated more heat during thread execution than others. It could be shown that cooperative thread migrations can significantly improve the thermal balance on chip. The heuristics that can measure the temperature outperformed the heuristics that do not possess this knowledge. Moreover the heuristics that base their mapping decision entirely on the learned heat sources of a (thread,core) tuples lower the overall temperature profile of the chip while only performing few thread migrations compared to the other cooperative heuristics. For the eight-core layout, the heuristics that take into account the different core shapes lower the overall temperature of the chip compared to the uninformed strategies (such as the naive, rotating and random strategies) for tge second scenario. Hence, to obtain the best thermal management the system should monitor the temperature distribution and should learn the heat sources of all (thread,core) tuples, i.e., by using the proposed thermal model.

The next chapter, will introduce the novel concept of self-aware computing and will give an outlook on how self-adaptive multi-cores can be extended towards self-awareness. Furthermore, a road map towards a self-aware hybrid multi-core will be discussed.

   I wonder if I've been changed in the night? Let me think:
was I the same when I got up this morning? I almost think
I can remember feeling a little different. But if I'm not the
same, the next question is, Who in the world am I? Ah,
THAT'S the great puzzle!

Alice, *Alice's Adventures In Wonderland*

CHAPTER 8

Making the Case for Self-aware Hybrid Multi-cores

Computing systems will become parallel and heterogeneous in order to keep up with the ever-increasing performance goals while lowering the power consumption at the same time. Ambitious application domains demand for a wide range of requirements such as functionality, flexibility, performance, resource usage, costs efficiency, reliability, safety, and security. As the applications change their requirements at run-time, the architecture of future computing systems must be able to adapt to the changing requirements of the applications.

Design-time solutions will no longer be able to provide the flexibility to provide acceptable solutions for all possible requirements without providing an enormous amount of system resources. Reconfigurable computing systems offer a less expensive solution in order to provide various levels of application requirements. When the requirements change, the system architecture can be reconfigured in a way such that the application's requirements can be fulfilled. Next to application requirements, the computing system has to adhere to dynamic environmental conditions such as the battery level (for embedded devices) and the chip temperature.

Handling the changing application requirements and environmental conditions concurrently at run-time imposes great challenges to the programmers of such systems. In computer science it is common to take inspiration from other sciences, i.e. natural sciences, in order to develop novel techniques that outperform state-of-the-art solutions. For instance, artificial neural networks are used for image classifications [40] and artificial ant colonies for solving the traveling salesman problem [43]. Recently, the idea of making computing systems 'self-aware' attracted a great deal of attention in the academic field [20, 27, 88]. Self-aware computing aims to handle various dynamically changing internal and external goals autonomously in a highly dynamic environment. Hence, introducing self-awareness into computing systems such as hybrid multi-cores sees to be a promising way to solve the mentioned challenges.

This chapter first gives a brief background on the origins of self-awareness in psychology and medicine. Then, current approaches of self-aware computing systems are presented. Finally, an outlook towards a self-aware hybrid multi-core is given.

8.1 Background on Self-awareness

The term self-awareness was first introduced in cognitive science around the start of the 20th century [25, 143]. Starting from the 1960s, many publications about self-awareness were published in the field of psychology. Duval and Wicklund proposed the self-awareness theory, which states that at any given time a person can either direct his attention to the environment or to himself, but he can not do both at the same time. Self-awareness has then been described as the ability of a person to direct his attention towards himself and to judge whether his behavior meets his personal goals, standards and values or not [45]. Duval and Wicklund [45] along with Goukens [59] divided the self-awareness into a public/subjective self-awareness and a private/objective self-awareness.

Duval and Wicklund defined private/objective self-awareness as follows:

> "When attention is directed inward and the individual's consciousness is focused on himself, he is the object of his own consciousness–hence 'objective' self awareness." [45]

Feeling hungry or feeling pain are prime examples for private self-awareness as this knowledge cannot be observed by another individual. An observer might notice that you are in pain, but he cannot tell how much pain you feel and where you feel the pain unless you tell him.

In contrast to that, for public/subjective self-awareness, the individual directs his attention away from himself and

> "experiences himself as the source of perception and action." [45]

Public self-awareness covers aspects that go beyond the single individual such as social relationships. Therefore, public self-awareness requires knowledge about the environment and other individuals. A common question for public self-awareness is 'how do others see me?'. Note that public and private self-awareness might stand in conflict when there are discrepancies between the social expectations of a group and the individual's values [50]. This conflict might be personally experienced as peer pressure.

The most famous self-awareness experiment is the mirror test, where a person or an animal is declared self-aware if he/it is able to identify himself/itself in the mirror. Carver and Scheier [35] have shown in experiments with female undergraduates that placing a mirror into a room does manipulate the undergraduates' response to a sentence completion blank. The authors concluded that the mirror increased the self-attention. In a second experiment, an audience was used instead of the mirror and it was shown that the audience further increased the self-attention.

Froming et al. [50], however, found out that the mirror and the audience have different effects on the self-awareness of a person. In their experiments the test person plays the role of a teacher who punishes his student for incorrect answers. The test persons were asked in preparation of the experiment if they approve or oppose the use of punishment in learning. It could be shown that a mirror manipulated the test persons to select a degree of punishment that reflects their personal attitude towards punishments. However, when facing an audience, the test persons changed their level of punishments towards the expectations of the audience. Hence, the authors concluded that the presence of a mirror increases the private self-awareness while the presence of an audience increases the public self-awareness.

Rochat [118] divided the self-awareness into five different levels, namely differentiation, situation, identification, permanence and meta self-awareness. Rochat studied at which ages these levels develop in childhood. The first level differentiation unfolds already at birth were newborns already seem to be able to sense their body as a differentiated entity. After two months, the infants can additionally sense how their body is situated in relation to other entities. At the age of approximately 14-18 months, children start to perceive traces of themselves and the so called 'me' is born, which gets extended over the fourth level. At the end of the fourth level, at the age of four, children understand that their self has a temporal dimension and they can recognize themselves on older photographs. By 4-5 years children start to wonder what others might perceive about them and develop more advanced and multiple representations and perspectives about themselves. Similar to Rochat, Morin [103] defines different level of consciousness and self-awareness of different complexity.

In summary an increased self-awareness helps a person to either meet his internal goals, values, and beliefs and/or the expectations of the environment. The question is now, how self-awareness can be defined for computing systems and how self-awareness can be directed or controlled within such systems. By taking inspiration from cognitive science, self-awareness might be split into a private part and a public part. Furthermore, different levels of self-awareness might need to be considered. The next section will first differentiate self-awareness from other self-* properties, which have been emerged in computing systems over the last years. Then it will present current approaches to translate self-awareness from cognitive science to computing systems.

8.2 Self-aware Computing Systems

In the last decade several so-called self-* properties have been proposed, which refer to the capability of a system to change itself autonomously without any external control in reaction to or anticipation of system dynamics. Among the research fields that employ self-* properties are `self-organizing` systems, `multi-agent` systems, `autonomic` computing, and `organic` computing. According to Di Marzo Serugendo et al. [125], `self-organizing systems` can change their internal structure and functionality at run-time without any explicit direction mechanism.

`Multi-agent systems` are one possible solution for designing artificial self-organizing systems by using several autonomous software agents that make local decisions and interact with each other to achieve their goals [151]. Multi-agent systems distinguish between two types of environment: a physical and a social environment. The environment can be characterized as accessible/inaccessible, deterministic/non-deterministic, discrete/continuous and static/dynamic.

The `autonomic computing concepts` aims to solve the emerging complexity crisis in software engineering where software engineers are no longer able to deal with the rising complexity, dynamic, heterogeneity,of future systems. Hence, future (autonomic) systems should be enabled to manage themselves. This self-management includes self-configuration, self-optimization, self-healing and self-protection. To enable autonomic computing four additional attributes were proposed: `self-awareness`, `environment-awareness`, `self-monitoring` and `self-adjustment` [140].

In autonomous computing systems, all system components are autonomic themselves and an autonomic manager is in charge of monitoring the components and the environment and in charge of developing and executing plans based on the analysis of this information. This approach is closely related to a hierarchical multi-agent system. IBM, who as a main driving force behind the autonomic computing idea, proposed the a reference architecture named `MAPE-k` for an autonomic manager, which operates the monitor, analyze, plan, and execute

(MAPE) control loop and maintains a knowledge base [74]. The architecture uses sensors to collect information about the environment and the system itself. In organic computing, the systems are not supposed to be fully autonomous. Hence, an observer and a controller component are introduced, which analyze, if the organic system (under observation) meets the system's goals and if not, interferes [124]. In order to make an organic system observable it requires an additional self-explaining property where the system can explain its current structure and behavior to either the observer or to an external user. External users can provide goals to the controller. Only, when the organic system does violate these goals or any other given constraints, the controller interferes. The underlying architecture for the organic system can be centralized, hierarchical, or decentralized. Self-awareness is a key attribute in both autonomic and organic computing. Furthermore, it can be seen that self-awareness can be divided into different properties such as environment-awareness, self-monitoring and self-adjustment/adaptation.

The vision of introducing self-awareness into computing systems was already presented by the Defense Advanced Research Projects Agency (DARPA) of the United States government in 2003 [112]. Six years later the European Commission launched a framework program for future and emerging technologies with the title 'Self-Awareness in Autonomic Systems (AWARENESS)' and currently funds and supports four different projects. These interdisciplinary projects investigate the question: "What is self-awareness in autonomic systems?" [17].

In the last years various research projects worked on definitions for self-awareness in computing systems and engineered first prototypes. Agarwal [20], Hoffmann [67, 68], Santambrogio [26, 122] and Sironi [132–134] are among the pioneers in self-aware computing. Agarwal et al. defined the following five properties for a self-aware computer:

1. "It is INTROSPECTIVE or SELF-AWARE in that it can observe itself and optimize its behavior to meet its goals.

2. It is ADAPTIVE in that it observes the application behavior and adapts itself to optimize appropriate application metrics such as performance, power, or fault tolerance.

3. It is SELF HEALING in that it constantly monitors its resources for faults and takes corrective action as needed. Self healing can be viewed as an extremely important instance of self awareness and adaptivity.

4. It is GOAL ORIENTED in that it attempts to meet a user's or application's goals while optimizing constraints of interest.

5. It is APPROXIMATE in that it uses the least amount of precision to accomplish a given task. A self-aware computer can choose automatically between a range of representations to optimize execution – from analog, to single bits, to 64-bit words, to floating point, to multi-level logic." [20]

Recently, Lewis et al. [88] came up with a novel approach that combines self-awareness and self-expression inside proprioceptive compute nodes. In contrast to previous definitions, Lewis et al. have divided the self-awareness into a public and private part, which is in line with the definition of self-awareness in neurocognitive science [45, 59]. Lewis et al. define a self-aware compute node as follows:

"To be self-aware a node must:

- Possess information about its internal state (private self-awareness).

- Possess sufficient knowledge of its environment to determine how it is perceived by other parts of the system (public self-awareness).

Optionally, it might also:

- Possess knowledge of its role or importance within the wider system.

- Possess knowledge about the likely effect of potential future actions / decisions.

- Possess historical knowledge.

- Select what is relevant knowledge and what is not." [88]

Finally, Lewis et al. set another focus on the process of finding an appropriate action to adapt the system according to the knowledge, which has been obtained by self-awareness. This process is called 'self-expression' and is again inspired by cognitive science. The Oxford English dictionary defines self-expression as the "expression of one's feelings, thoughts or ideas especially in writing, art, music, or dance" [18]. According to neurocognitive science, persons are more likely to express themselves when their actions are in line with their states and traits [37]. Lewis et al. define a self-expressive compute node as follows:

- "A node exhibits self-expression if it is able to assert its behaviour upon either itself or other nodes.

- This behaviour is based upon the node's state, context, goals, values, objectives and constraints." [88]

Besides working definitions for self-awareness, several prototypes for self-aware computing systems were published recently. The following related work used the self-awareness working definitions of Agarwal et al. [20] and applied them to embedded systems, multi-core processors, workstations with an FPGA accelerator board, and exascale computers. Self-awareness is provided using an observe-decide-act control loop, which is depicted in Figure 8.1.

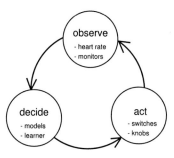

Figure 8.1: Observe-decide-act control loop (source: [20])

Performance, power consumption and further metrics of an application are observed using monitors. For monitoring the performance, related work mostly applies the Heartbeats framework [67], which was already described in Section 2.4. In the decision phase, different performance models and learners can be used to select an appropriate adaptation whenever the system or application goals are not met. In the action phase, different switches can be modified or knobs can be turned in order to adapt the system in such a way that the corresponding goals will be fulfilled in the future. The switches and knobs represent the different adaptation techniques available in the system. The switches might represent different implementation types of an application's task and the knobs might represent further system parameters, such as the clocking rates and voltages of the cores.

Sironi et al. [134] presented an FPGA-based self-aware adaptive computing system, which includes performance monitoring, decision making and self-adaption. To monitor the performance of an application, the Heartbeats framework was used. To adapt the performance, the application could either be mapped in software, in hardware or both. An encryption algorithm was used as application and was implemented on a Xilinx Virtex-II Pro FPGA. However, the experimental results did only cover static measurements of the application's performance and provided self-adaptation only conceptually.

Looking at computer systems in general, Santambrogio et al. [122] proposed enabling technologies for self-aware adaptive systems. Again the Heartbeats framework was used to monitor the application's performance and self-awareness was achieved using the observe-decide-act loop. The authors introduced a novel decision making process named 'Smartlocks'. Smartlocks is a self-optimizing spin-lock library, which includes an adaptable lock acquisition scheduling policy. In order to observe and optimize an application's performance, Smartlock uses the Heartbeats framework combined with a machine learning engine.

In [133], Sironi et al. discussed a heterogeneous system that consists of a multi-core general purpose processor and a reconfigurable device. For experimental evaluation, an Intel Core i7 and a Xilinx Virtex-5 FPGA were used as a prototyping platform. The authors show the efficiency of their approach in an experiment with four instances of an application, which hashes data blocks with given performance goals. Using a hot-swap mechanism that switches between a software implementation on the processor and a hardware implementation on the FPGA, all four heart rates could be met. The authors, furthermore, emulated system dynamics, such as a utilizing a varying subset of system resources, and demonstrated that the proposed self-aware adaptive methodology is able to handle these dynamics.

Recently, Sironi et al. [132] proposed Metronome, a framework that extends operating systems with self-adaptive capabilities. In particular the observe-decide-act control loop is implemented by a heart rate monitor (observe) and a performance-aware fair scheduler (decide, act). The target architectures for the Metronome framework are multi-core processors and the host operating system is Linux. Using the proposed scheduler a task that does not meet its performance goal can get assigned more execution time. For evaluation the Linux scheduling infrastructure was extended and experimental results were performed for the PARSEC[24] 2.1 benchmark on a workstation with a single Intel Core i7-870 quad-core processor running at 2.97 GHz. The results showed that the proposed scheduler is able to reach the desired performance goals for two concurrently executing benchmark applications. In contrast, the completely-fair scheduler of Linux provided a higher performance than necessary for one application and a too low performance for the second application.

Targeting exascale computing, Hoffmann et al. [68] presented the conceptual Angstrom processor, which is a many-core processor targeting up to 1,000 cores on a single chip. The Angstrom processor provides architectural support for self-aware computing. Therefore, the paper discussed several hardware extensions such as special performance counters (observe), cache adaptations, and special low-power cores (act). Simulation results demonstrated that a self-aware computing

---

[24]Princeton application repository for shared-memory computers (PARSEC)

system clearly outperformed a non-adaptive system in the metric performance per watt.

Similar to this thesis, Bartolini et al. [26] addressed performance and thermal management on a multi-core. In the paper, a framework is presented that inserts idle cycles into a thread to control the chip's temperature as long as the quality of service constraints and service level agreements are still met. Furthermore, the framework can adapt the thread priorities in order to reschedule threads such that they meet their service level agreements. For experimental evaluation, applications of the PARSEC 2.1 benchmark are studied. The authors compared their results to a related thermal management technique called 'Dimetrodon' that also inserts idle cycles in order to reduce the temperature of the chip. In contrast to Dimetrodon, the performance of the studied benchmarks could be improved between 12% and 57% because the proposed technique inserted less idle cycles while achieving a comparable chip temperature.

In another experiment, a four-threaded benchmark application named 'swaptions' could be kept inside a predefined performance interval while keeping the chip temperature around 60°C. This work already combined performance and thermal management, which is a desirable next step for the presented methodologies of this thesis as well. However, the authors targeted homogeneous multi-processors and not hybrid multi-cores, which also contain reconfigurable hardware cores. Similar to Bartolini et al., we believe that by applying self-aware methodologies systems can efficiently and autonomously find dynamic trade-offs between concurrent system and application goals, such as performance intervals of applications and peak temperature of the chip, at run-time.

8.3 Outlook on Self-aware Hybrid Multi-cores

One of the integrated projects of the AWARENESS initiative funded by the European Commission is 'EPiCS: Engineering Proprioception in Computing Systems' [19]. The EPiCS project seeks to find a common reference architecture framework that applies to several computer architectures and various application domains. As one result of the EPiCS project, Lewis et al. [88] defined a novel working definition for proprioceptive computing, which combines the concept of self-awareness with the concept of self-expression.

Figure 8.2 depicts the reference framework for a proprioceptive compute node, which has been defined as part of the EPiCS project and contains all fundamental components in order to enable self-awareness and self-expression. The framework is generic in the sense that it can be applied to any computer architecture. Since this framework supports and structures the development of a self-aware and self-

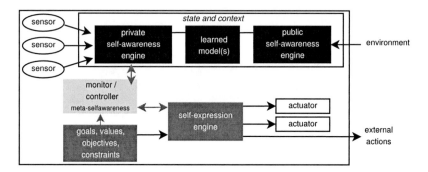

Figure 8.2: Reference architectural framework for a proprioceptive node (source: [27]).

expressive computing system, this section presents how the individual components of this framework can be implemented on a hybrid multi-core. Furthermore, more details will be given to answer the questions: 'What can a hybrid multi-core aware of?' and 'How can a hybrid multi-core express itself?'.

The proprioceptive compute node is composed of a `self-aware` subsystem, which captures and maintains the system state and context, and a `self-expressive` subsystem, which adapts the system's behavior to changing requirements. The self-aware subsystem is displayed in the upper part of Figure 8.2 by dark-gray boxes. According to [88], a self-aware computing system must posses knowledge of and based on phenomena internal and external to itself where the internal phenomena (system state) are handled by a `private self-awareness engine` and the external phenomena (system context) are handled by a `public self-awareness engine`. The system captures the system state using internal sensors and the system context by observing the environment and the interaction to other (proprioceptive) nodes. Both self-awareness engines use their input to continuously learn and update models of the system itself and conceptual knowledge of the environment of which the system is part of.

The self-expressive subsystem (medium-gray boxes) takes the system state and context as input in order to adapt the system's behavior whenever the system does not meet its goals, values, objectives, or constraints. To achieve this, a `self-expression engine` contains a single strategy or a set of different strategies. The self-expression is realized by using internal actuators and/or by taking external actions. For instance, on a hybrid multi-core the actuators could migrate threads between different cores in order to increase an application's performance, to reduce the system's power consumption, or to balance the on-chip temperature distribution. In a same vein, migrating a thread to another proprioceptive node is an example for an external action. The system's goals, values, objectives, and constraints might be given at design-time or might be dynamically received/updated at run-time from another node or an external user.

Optionally, a proprioceptive compute node can implement `meta-self-awareness`, which can be seen as a second-level self-awareness where the system is aware that it is aware. Here, a `monitor/controller` component (light-gray box) observes how well the self-aware subsystem and the self-expressive subsystem achieve the system's goals, values, objectives, and constraints. The meta-self-awareness is mostly used to deal with trade-offs of conflicting goals, which might change at run-time depending on the system dynamics. Then, the controller might decide that the self-expression engine should perform a different strategy or that a self-awareness engine should use a different subset of sensors or should employ a different on-line learning technique. The monitor/controller component is inspired by the observer and the controller components of an organic computing system.

In the following, we will describe how the reference architectural framework can be applied to a hybrid multi-core. The internal sensors can be diverse, i.e. temperature sensors, performance counters, power sensors, and further monitors that capture the core utilization, number of cache misses, etc. These sensor readings can not be directly accessed by other nodes and represent the internal system state. Limiting the maximum overall temperature of the chip or the spatial temperature differences can be a private system goal just as minimizing the power consumption. Another internal goal might be a uniform core utilization over time in order to limit the physical wear-out of cores. Users that want to execute applications on this node might not be interested in any of these goals. Therefore, self-aware hybrid multi-cores should autonomously handle these goals without user interaction to protect the health of the system.

As input for the public self-awareness engine, the hybrid compute node receives input data for the applications from external devices such as the video input port or the Ethernet port (using TCP/IP packets). Furthermore, user-defined performance, power, and thermal goals/objectives/constraints for specific applications are part of the environment as well. Finally, the self-aware hybrid multi-core might be part of a network of self-aware and/or non-self-aware nodes. In this case, the compute node might improve its behavior by learning its role inside the network. For instance, let the location of a node A in the network be a connecting point between different subnetworks. Then A's most important role in the network is to play a hub for the different subnetworks. Then, this node might reject to execute any other task and adapt itself in a way that it can forward data packets between the different subnetworks at the highest possible rate. In highly dynamic networks the topologies can change over time, thus, knowing the current role in the network can be key for an efficient execution.

As a basis for public and private self-awareness, the proprioceptive node requires models for the system state (itself) and for the context it is placed in (environment, network of nodes). If the system state or the context change in such a way that the models become inaccurate, the self-aware node should apply machine learning autonomously in order to update the affected models. Considering thermal management, the thermal model of the chip has to be updated when the fan on the chip is activated/deactivated because this dramatically influences the heat flow on the chip. In this example, the system can observe that the thermal model of the chip has become inaccurate by the growing error of thermal predictions.

Whether the models are still accurate or not and whether the applied learning method is sufficient can be controlled by an additional monitor that provides meta-self-awareness. The reference architectural framework provides this second-level awareness by monitoring how well the public and private self-awareness engines fulfill their requirements. The meta-self-awareness can also be used to handle conflicts between concurrent private and public goals (or objectives).

System adaptations are performed by the system autonomously either when a goal, objective, value, or constraint is not met or a metric could be optimized towards a goal or an objective. For self-adaptive hybrid multi-cores, this thesis has proposed to apply intra-modal and trans-modal thread migrations. This level of adaptiveness might not be sufficient for fine-granular system adaptation. Hence, future self-aware hybrid multi-cores might support further adaptations such as dynamic frequency and voltage scaling, idle cycle insertions [26] or self-optimization of specific thread implementations.

For instance, Becker et al. [28] proposed a system that replaced state-of-the-art hardware multipliers with fixed-coefficient multipliers where they reconfigured the fixed-coefficient whenever the second coefficient of the multiplication changes. This approach makes sense when the input for second coefficient stays constant over long time periods. Then, the proposed self-optimization significantly improves performance while reducing the area requirements and power consumption at the same time.

Moreover, external actions such as horizontally migrating threads between compute nodes might be possible for future networks of self-aware compute nodes. For a hardware thread, a migration might work when both hybrid multi-cores are implemented on identical platform FPGAs and when both nodes have the same multi-core layout. However, for hybrid nodes this task is more challenging and maybe impractical as the thread would need to be recompiled for the new multi-core architecture, which can be a computational intensive process.

The biggest challenge, however, is to handle the different trade-offs between concurrent application goals, system goals and the goals of other compute nodes. Here, a self-aware compute node might need to prioritize its goals. Finding the right priorities is most likely a part of the meta-self-awareness, which also has to decide if the priorities or preferences have to be updated over time.

Many great challenges have to be faced in order to create a self-aware system. One potential blueprint of a self-aware hybrid multi-core has been drawn in this section. This thesis has presented a self-adaptive hybrid multi-core, which already implements basic parts of a future self-aware hybrid multi-core. A timebase was used as performance counter and ring-oscillators have been used as temperature sensors. A video object tracking case study was performed on a hybrid multi-core respecting user-defined performance goals. Here, the self-adaptive system aimed to reduce the number of used cores, which can save power. Hence, the results for performance management can already be seen as a rudimentary self-aware system. In contrast to the performance management, the model for thermal management is learned at run-time. The proposed self-adaptation techniques for thermal management might be used later by a self-aware hybrid multi-core in order to balance the temperature distribution or possibly to lower the overall temperature profile of the chip.

8.4 Chapter Conclusion

This chapter gave an extended outlook towards self-aware hybrid multi-cores. In a first step, the chapter discussed the origins of the term self-awareness in psychology. In psychology, self-awareness is described as the ability of a human to direct his attention towards himself (and not towards the environment) in order to judge if his behavior meets his personal goals, objectives, and values. By realizing discrepancies between his past behavior and his personal values or goals, a person usually changes his behavior in favor of his values and goals. Scientific literature distinguishes between private and public self-awareness and between different levels of self-awareness.

In computer science, a growing number of researchers believe that by employing concepts of self-awareness future computing systems will be able to autonomously manage the rising complexities and run-time dynamics imposed by the applications, quality of service constraints, computer architectures, and their environments. For self-aware computing, two working definitions and first prototype implementations on various computing system architectures were described.

In a last step, the reference architectural framework of the EPiCS project was introduced and applied to a hybrid multi-core architecture. The chapter made several suggestions on how different features of the reference architectural framework can be realized on a hybrid multi-core. This chapter also discussed which parts of the framework were already realized by the self-adaptive hybrid multi-core as proposed in this thesis.

> *I quite agree with you, and the moral of that is—"Be what you would seem to be"—or if you'd like it put more simply—"Never imagine yourself not to be otherwise than what it might appear to others that what you were or might have been was not otherwise than what you had been would have appeared to them to be otherwise."*
>
> The Duchess, *Alice's Adventures In Wonderland*

CHAPTER 9

Conclusions and Future Research Directions

This chapter summarizes the contributions of this thesis and draws conclusions from its experimental and simulation results. Finally, the chapter gives an outlook to future research directions.

9.1 Contributions

Today's trends in embedded computer architectures indicate that future systems will (i) compose multiple processors on a single chip, will (ii) have an increased heterogeneity between the processors and will (iii) integrate more (reconfigurable) hardware accelerators in order to increase the performance while minimizing the power consumption at the same time. Hybrid multi-cores combine these three trends on a single chip.

By extending the well-known multithreading paradigm towards reconfigurable hardware, programming applications for hybrid multi-cores is comparable to programming applications for multi-processor systems. For most systems the functionality of reconfigurable hardware threads has still to be programmed using a hardware description language. In ReconOS the communication and synchronization between hardware and software threads is handled transparently by the operating system using the well-known POSIX primitives. Also, the shared memory can be easily accessed using dedicated interfaces to the memory.

However, while programming models and execution environments have been successfully developed in the last years, the run-time management of such systems is still a major research field. This thesis has set its focus on performance and thermal management on hybrid multi-cores with the goal that the system can autonomously fulfill user-defined performance goals of applications and can autonomously minimize spatial temperature differences on chip, i.e. by avoiding local temperature hot spots, using self-adaptive thread re-mapping. For self-adaptive thread mapping, the system captures the system state, such as the performance of the application/s or the temperature distribution of the chip, using monitoring cores, such as performance counters and temperature sensors.

Most related work addresses multi-processor systems for implementing thread mapping strategies. This thesis provides fundamental novel results for the run-time management of self-adaptive hybrid multi-cores focusing on performance and temperature. This work provides the following contributions:

- We have implemented prototypes of self-adaptive hybrid multi-cores on modern FPGAs using the programming model and execution environment ReconOS. Ring oscillator-based temperature sensors and classic timers were used as monitors for capturing the on-chip temperature distribution and the performance of applications, respectively. The temperature sensors were calibrated using a novel self-calibration technique that relies on regional heat-generating cores and a built-in thermal diode. The prototype allowed intra-modal and trans-modal thread migrations. The thread migrations were, however, emulated using dynamic creations and terminations of threads.

- For the implementation of the prototypes on modern FPGAs it was necessary to develop a new version of ReconOS, which takes into account the recent changes in the FPGA technologies and the corresponding tool flows. Therefore, we have contributed to the ReconOS version 3.0, which uses soft-core processors instead of hard-core processors, has a newly-designed operating system interface, and a new memory subsystem for reconfigurable hardware cores.

- For performance management, models and self-adaptation algorithms have been developed for parallel streaming applications and experimentally evaluated on a real-world case study. The self-adaptation techniques either aim at providing user-defined quality-of-service constraints, i.e., a lower performance bound or a user-defined performance interval, while minimizing the number of used cores at the same time. The experimental evaluation of the proposed self-adaptation algorithms demonstrated the effectiveness on a multithreaded video object tracking case study.

- For thermal management, we have first demonstrated that significant spatial temperature differences can already be created and measured on today's FPGA chips. Therefore we have developed a prototype with a set of regional heat-generating cores and a regular grid of ring-oscillator based temperature sensor. Detailed design space explorations of dedicated heat-generating cores and sensor layouts were obtained as part of this thesis. Furthermore, a two-layer thermal model was proposed, which is based on an RC network similar to the temperature simulator HotSpot. Unlike HotSpot, the model parameters were learned at run-time using randomized hill climbing. The achieved experimental results demonstrated the high prediction accuracy of the thermal model. Next, the thesis has introduced novel thread mapping strategies in order to balance the chip temperature and possibly lowering the overall chip temperature. Eleven different non-cooperative and eleven cooperative strategies were evaluated for different scenarios and multi-core layouts using a simulator that is based on the learned thermal model of a Virtex-6 FPGA. The benefits of allowing cooperative thread migrations at defined time steps during thread execution and the benefits of learning the heat sources of the threads using the thermal model were quantified for different scenarios.

The proposed self-adaptive hybrid multi-core sets a basis for future self-aware compute nodes. A self-aware computing systems has to be able to capture and maintain its state and context. Here, the system employs local sensor information, e.g., the temperature readings of a thermal diode, to observe the current system state. The environmental context can be captured by observing the interaction with other compute nodes, the input data for applications, and the user-defined quality-of-service constraints. The self-adaptive hybrid multi-core prototype already provides internal sensors such as performance counters and temperature sensors, which can be used as input for private self-awareness. The proposed thread mapping strategies are among the possible self-expression techniques, which can be applied by the system in order to fulfill its internal goals, objectives, constraints, and/or its external quality-of-service constraints. In future, conflicting private and/or public goals or constraints have to be handled using a second-level awareness (meta-self-awareness). Furthermore, the interaction of a self-aware hybrid multi-core inside a network of self-aware nodes opens a novel field in research.

The developed prototypes of our self-adaptive hybrid multi-core already serve as a basis for diverse research directions in self-aware computing in the context of the EPiCS collaborative research project. Our hybrid multi-core architecture is used to implement prototypes for novel adaptive protocol stacks [78–80] at the Eidgenössische Technische Hochschule Zürich (ETHZ), Switzerland, and is furthermore used as the target architecture for providing fault-tolerance and safety using dynamic thread redundancies at the European Aeronautic Defence and Space Company N.V. (EADS) in Munich [27]. Finally, the manually defined thermal management techniques and the temperature simulator are used as a starting point for learning new mapping heuristics autonomously on the hybrid multi-core at run-time. This research is investigated in cooperation with the University of Birmingham in the United Kingdom as part of the EPiCS project.

9.2 Conclusions

Based on our experimental evaluations, we draw the conclusion that self-adaptive hybrid multi-cores can autonomously perform run-time management, such as performance management and thermal management, using run-time adaptations at thread level. We have investigated an video object tracker as an example for a complex real-world case study and have demonstrated how autonomous performance management can be applied on such a system. We believe that our results also hold for other parallel streaming applications such as audio and video processing or data encryption/decryption. Our performance models and algorithms hold for any kind of application that posses threads that can be instantiated several times in order to divide the workload between the thread instances. Furthermore, we conclude that a control loop of monitoring the execution time of the hybrid threads, deciding for possible thread re-mappings, and finally adapting the thread mapping is a powerful technique to not only react to significant fluctuations in the application's performance but also to minimize the core utilization.

Furthermore, we have proven that we can use current FPGA devices to emulate the thermal challenges of future embedded systems. Using dedicated regional heat cores we could generate a temperature increase of up to $134°C$ over twelve minutes. Next to temporal temperature increases, we could also generate and measure spatial temperature differences on a chip of up to $6.5°C$ over the entire chip. Hence, the results of this thesis enable future researchers to evaluate their thermal management techniques not only in simulation but also using experiments on a prototype. Our novel self-calibration technique for ring oscillator-based temperature sensors can also be used for many further research directions and also industrial projects, which require additional sensors in their designs. Using self-calibration, the extensive manual calibration using expensive external devices is replaced by a fully-automatic internal calibration, which significantly reduces the financial costs and the overhead for sensor calibration.

By learning the model parameters at run-time, we were able to predict the on-chip temperature distributions with an accuracy of $0.72°C$ on average. We believe that this accuracy is adequate to employ the thermal model to predict the effects of updating the thread mapping at run-time. Furthermore, by applying tour thermal model the thermal footprints of threads can be learned at run-time after the static model parameters have been learned in an initial phase. As an example for a thread, we have learned the heat sources of a specific regional heater in our experiments.

Moreover, we conclude that by smartly adapting the thread mapping at run-time, the hybrid multi-core can balance the chip temperature. We have shown that by taking into account the current core temperatures for thread re-mapping, the thermal balance of the chip can be improved. Allowing cooperative thread migrations between the hybrid cores shows better results for the thermal balance on chip. By taking advantage of the knowledge of the thermal footprints of (thread, core) tuples, the numbers of costly thread migrations can be decreased by an order of magnitude with the result that the temperature profile of the chip can be reduced while the thermal balance does not improve. We conclude that our heuristics can be used to meet different optimization problems. However, we assume that there is more space for even smarter self-adaptation strategies.

This thesis has focused on performance management and thermal management separately. Other run-time management techniques such as dynamic power management were not covered. Nevertheless, we believe that further run-time management techniques can be implemented on hybrid multi-cores as well using the same concept. Our concept requires a monitor, which measures the system state, a system model, which estimates the effects of thread re-mappings, and an algorithm, which selects the appropriate thread re-mapping. For instance, our concept could be applied to dynamic power management. Here, a power-meter could be used to monitor the overall dynamic power consumption of the entire chip (or the power consumption of certain FPGA regions) and a power model could assign power consumption values to (thread, core) tuples. Then, various self-adaptation strategies can intend to meet various user-defined power goals. Therefore. we believe that certain heuristics, which were developed for performance or thermal management, can be reapplied with minimal changes.

9.3 Outlook

We see the following major directions of future research, which are related to our proposed self-adaptive hybrid multi-cores and the corresponding run-time management techniques:

Trans-modal thread migration: Thread migrations across the HW/SW boundary impose various challenges since the parallel processing on (reconfigurable) hardware logic is fundamentally different to the corresponding sequential processing on a processor. Thus, the thread context is usually dissimilar for the hardware and software implementations. As part of this thesis it was already suggested to extend the cooperative multithreading approach of ReconOS [94] from hardware threads towards hybrid threads. Hybrid threads can run in software and in reconfigurable hardware and can be dynamically migrated between both modalities. In cooperative multithreading the threads have explicitly defined migration points. Whenever a thread reaches such a point it informs the operating system that a thread migration can take place. However, these migration points are usually defined manually [58]. When the applications are specified using synchronous data flow graphs or Kahn networks, convenient migration points might be extracted automatically where the thread context is minimal and can be translated between hardware and software. There is already initial research for an automatic context translation by applying a common task specification, which is then implemented in hardware and software [82]. However, only tiny programs were studied. Whether the proposed morphing approach can also be applied to complex applications is still an open research question.

Self-aware hybrid multi-cores: Applying further concepts of self-aware computing on the self-adaptive hybrid multi-core is another interesting research direction. One challenge is the autonomous handling of conflicting public and private goals/constraints [88]. Introducing a novel meta-self-awareness layer that observes and controls the private and public self-awareness engines might be a promising approach to deal with these conflicts. For instance, this layer can switch between different the self-adaptation/self-expression strategies in order to find an adequate trade-off between conflicting goals for the current scenario. More system goals, e.g., minimizing the system's power consumption, can be managed when further system monitors such as a power-meter and further levels of adaptivity, such as dynamic frequency and voltage scaling or the dynamic integration of idle cycles into a thread's execution, are supported by the system.

Safety-critical systems: The self-adaptive hybrid multi-core is a promising architecture for safety critical systems, such as aircraft electronics, because dynamic thread redundancies can be achieved by instantiating multiple redundant copies of critical threads. We assume that the criticality of a thread depends on the current state of the system. For instance, when an orbiter has to correct its flight path the thread that is responsible for the navigation becomes critical. At another time, when the orbiter takes pictures and analyzes them, different threads are important. Hybrid multi-cores can reduce the area requirements for the chip and therefore the costs because the redundant thread instances can be dynamically instantiated and (re)configured to the system. The most interesting research challenges are the integration of the dynamic thread redundancy, check-pointing, and thread recovery techniques into a hybrid multi-core. Furthermore, it might be worthwhile to consider thermal aspects such as the core temperature for the mapping of critical threads. Some of these aspects are investigated by EADS as part of the EPiCS project, where our self-adaptive hybrid multi-core defines the basis for investigating dynamic fault-tolerance [27].

Autonomic networking: The self-adaptive/self-aware hybrid multi-core might be a compute node inside a wider network. In this case, applications might be spread between multiple nodes inside the network. Thus, one research challenge is the creation of an autonomous networking architecture. The networking architecture can be enabled to flexibly adapt the networking functionality according to the available system resources and the network traffic by using run-time reconfiguration. Furthermore, an exciting further research domain is the thread migration between compute nodes (horizontal migration). This is especially challenging for hybrid multi-cores, because different reconfigurable architectures (core layouts, FPGA packages) need to be considered for migrating hardware/software threads. Our self-adaptive hybrid multi-core architecture can be used to implement prototyping platforms for autonomous networking architecture. Indeed, the Communications and Systems Group of the ETHZ builds their prototypes on top of our self-adaptive hybrid multi-core [27, 80].

High-performance computing: While this thesis targets embedded system-on-chips, the proposed concepts can be also applied in the field of high-performance computing. Nowadays, compute clusters become more heterogeneous, e.g. by integrating special vector processors, general purpose graphical processing units, and FPGAs as accelerators. However, extending the multithreading approach to heterogeneous compute clusters implicates novel challenges such as resource sharing, thread synchronization, and communication.

In summary, we are confident that the proposed self-adaptive hybrid multi-core is a preview towards future (embedded) computing architectures. In the next decades the computer architectures, applications, user requirements, and run-time dynamics become so challenging, such that novel run-time management techniques will become necessary to fulfill all requirements. The research results of this thesis laid the foundations for further more advanced run-time management techniques such as self-aware computing.

List of Figures

List of Tables

List of Source Codes

List of Algorithms

Acronyms

API	application programmer interface
ASIC	application-specific integrated circuit
ATP	adaptive thread partitioning [mapping algorithm]
BRAM	block RAM
CDF	cumulative probability distribution function
CLB	configurable logic block
CPU	central processing unit
DARPA	defense advanced research projects agency
DCM	digital clock manager
DCR	device control register [bus]
DDR-SDRAM	double data rate synchronous DRAM
DFS	dynamic frequency scaling
DRAM	dynamic RAM
DSE	design space exploration
DSP	digital signal processor
DTM	dynamic thermal management
DVFS	dynamic voltage and frequency scaling
DVS	dynamic voltage scaling
EDK	embedded development kit [tool]
eDRAM	embedded DRAM
EPiCS	Engineering Proprioception in Computing Systems
FF	flip-flop
FIFO	first-in first-out [storage element]
FPGA	field-programmable gate array
FSL	fast simplex link
FSM	finite-state machine
GCC	GNU compiler collection
GUI	graphical user interface
HDL	hardware description language
HSV	hue / saturation / value [color space]
HW	hardware
I	importance [stage]
IC	integrated circuit
ICAP	internal configuration access port
IOB	input/output blocks
JTAG	joint test action group [debug interface]

LUT	look-up table
MEMIF	memory interface
MMU	memory management unit
MPEG	moving picture experts group [compression method]
MPSoC	multiprocessor SoC
NBTI	negative bias temperature instability
NoC	network-on-chip
O	observation [stage]
OS	operating system
OSIF	operating system interface
OSS	operating system service
PARSEC	Princeton application repository for shared-memory computers
PC	personal computer
PCB	printed circuit board
PCIe	peripheral component interconnect express [interface]
PF	particle filter [method]
PLB	processor local bus
POSIX	portable operating system interface
QoS	quality of service [constraint]
R	resampling [stage]
RAM	random-access memory
RC	resistor-capacitor [network]
RISC	reduced instruction set computing
RISP	reconfigurable instruction set processor
RMSC	root mean square error
rSoC	reconfigurable SoC
RSR	residual systematic resampling
RTL	register-transfer level
S	sampling [stage]
SIR	sampling importance resampling [algorithm]
SIS	sequential importance sampling [algorithm]
SMC	sequential Monte Carlo [method]
SMT	simultaneous multithreading
SoC	system-on-chip
SPEC	standard performance evaluation corporation [benchmark]
SRAM	static RAM
SRL	shift register LUT
SW	software
TCP/IP	Transmission Control Protocol / Internet Protocol
TLB	translation look-aside buffer
UART	universal asynchronous receiver transmitter [interface]
USB	universal serial bus
VHDL	VHSIC hardware description language
VHSIC	very high speed integrated circuit
VLSI	very-large-scale integration

Bibliography

Author's Publications

[1] Markus Happe. Parallelisierung und Hardware-/Software-Codesign von Partikelfiltern. Master's thesis, University of Paderborn, July 2008.

[2] Marco Platzner, Sven Döhre, Markus Happe, Tobias Kenter, Ulf Lorenz, Tobias Schumacher, Andre Send, and Alexander Warkentin. The GOmputer: Accelerating GO with FPGAs. In *Int. Conf. on Engineering of Reconfigurable Systems and Algorithms (ERSA)*. CSREA Press, 2008.

[3] Markus Happe, Enno Lübbers, and Marco Platzner. A Multithreaded Framework for Sequential Monte Carlo Methods on CPU/FPGA Platforms. In *5th International Workshop on Reconfigurable Computing: Architectures, Tools and Applications (ARC)*, pages 380–385. Springer, 2009.

[4] Markus Happe, Enno Lübbers, and Marco Platzner. An Adaptive Sequential Monte Carlo Framework with Runtime HW/SW Repartitioning. In *Int. Conf. on Field Programmable Technology (FPT)*. IEEE, Dec 2009.

[5] Jason Agron, David Andrews, Markus Happe, Enno Lübbers, and Marco Platzner. Multithreaded Programming of Reconfigurable Embedded Systems. *Reconfigurable Embedded Control Systems: Applications for Flexibility and Agility*, pages 31–54, IGI Global, 2010. ISBN: 978-16-096-0086-0.

[6] Markus Happe, Andreas Agne, and Christian Plessl. Measuring and Predicting Temperature Distributions on FPGAs at Run-Time. In *Proceedings of the International Conference on Reconfigurable Computing and FPGAs (ReConFig)*, Cancun, Mexico, December 2011. IEEE.

[7] Markus Happe and Enno Lübbers. A Hybrid Multi-Core Architecture for Real-Time Video Tracking. In *FPL 2011 Workshop on Computer Vision on Low-Power Reconfigurable Architectures*, Chania, Greece, September 2011. www.techfak.uni-bielefeld.de/~fwerner/fpl2011/abstracts/FPL11-CVW13.pdf.

[8] Markus Happe, Enno Lübbers, and Marco Platzner. A Self-adaptive Heterogeneous Multi-core Architecture for Embedded Real-time Video Object Tracking. *Journal of Real-Time Image Processing*, pages 1–16, 2011. 10.1007/s11554-011-0212-y.

[9] Markus Happe, Andreas Agne, Christian Plessl, and Marco Platzner. Hardware/Software Platform for Self-aware Compute Nodes. In *FPL 2012 Workshop on Self-Awareness in Reconfigurable Computing Systems*, Oslo, Norway, August 2012. http://srcs12.doc.ic.ac.uk/docs/paper_2.pdf.

[10] Markus Happe, Hendrik Hangmann, Andreas Agne, and Christian Plessl. Eight Ways to put your FPGA on Fire - A Systematic Study of Heat Generators. In *Proceedings of the International Conference on Reconfigurable Computing and FPGAs (ReConFig)*, Cancun, Mexico, December 2012. IEEE.

[11] Christian Plessl, Marco Platzner, Andreas Agne, Markus Happe, and Enno Lübbers. Programming Models for Reconfigurable Heterogeneous Multi-Cores. *Awareness Magazine, http://www.awareness-mag.eu*, 2012.

[12] Christoph Rüthing, Andreas Agne, Markus Happe, and Christian Plessl. Exploration of Ring Oscillator Design Space for Temperature Measurements on FPGAs. In *International Conference on Field Programmable Logic and Applications (FPL)*, Oslo, Norway, August 2012. IEEE.

[13] Markus Happe, Meyer auf der Heide, Peter Kling, Marco Platzner, and Christian Plessl. On-The-Fly Computing: A Novel Paradigm for Individualized IT Services (Invited Paper). In *Workshop on Software Technologies for Future Embedded and Ubiquitous Systems (SEUS)*, Paderborn, Germany, June 2013.

Bibliography

[14] Intel® Pentium 4 Processor in the 478-Pin Package Thermal Design Guidelines. http://download.intel.com/design/Pentium4/guides/24988903.pdf, May 2002.

[15] Intel 64 and IA-32 Architectures Software Developer's Manual Volume 3A: System Programming Guide, Part 1. http://www.intel.com/Assets/PDF/manual/253668.pdf, December 2009.

[16] *Organic Computing — A Paradigm Shift for Complex Systems Autonomic Systems*, volume 1. Springer Basel, 2011.

[17] FP7: FET Proactive Initiative: Self-Awareness in Autonomic Systems (AWARENESS). http://cordis.europa.eu/fp7/ict/fet-proactive/aware_en.html, Oct 2012.

[18] Oxford English dictionary. http://oxforddictionaries.com, Dec 2012.

[19] The EPiCS project: Engineering Proprioception in Computing Systems. http://www.epics-project.eu, Apr 2012.

[20] A. Agarwal, J. Miller, J. Eastep, D. Wentziaff, and H. Kasture. Self-aware computing. *Final Technical Report AFRL-RI-RS-TR-2009-161*, page 81, 2009.

[21] Andreas Agne, Enno Lübbers, and Marco Platzner. Memory Virtualization for Multithreaded Reconfigurable Hardware. In *IEEE International Conference on Field Programmable Logic and Applications (FPL)*. IEEE, 2011.

[22] Jason Agron and David Andrews. Distributed Hardware-Based Microkernels: Making Heterogeneous OS Functionality A System Primitive. In *18th Annual IEEE Symposium on Field-Programmable Custom Computing Machines (FCCM)*, 2006.

[23] M. Sanjeev Arulampalam, Simon Maskell, Neil Gordon, and Tim Clapp. A tutorial on particle filters for online nonlinear/non-gaussian bayesian tracking. *IEEE Transactions on Signal Processing*, 50(2):174–188, 2002.

[24] A. Athalye, M. Bolić, S. Hong, and P. M. Djuric. Generic Hardware Architectures for Sampling and Resampling in Particle Filters. *EURASIP Journal on Applied Signal Processing*, 2005.

[25] Smith Baker. The Identification of the Self. *Psychological Review*, 4(3):272–284, 1897.

[26] Davide Basilio Bartolini, Filippo Sironi, Martina Maggio, Riccardo Cattaneo, Donatella Sciuto, and Marco Domenico Santambrogio. A Framework for Thermal and Performance Management. In *Workshop on Managing Systems Automatically and Dynamically (MAD)*, 2012.

[27] Tobias Becker, Andreas Agne, Peter R. Lewis, Rami Bahsoon, Funmilade Faniyi, Lukas Esterle, Ariane Keller, Arjun Chandra, Alexander Refsum Jensenius, and Stephan C. Stilkerich. EPiCS: Engineering Proprioception in Computing Systems. In *Proceedings of the 10th IEEE/IFIP Conference on Embedded and Ubiquitous Computing (EUC) to appear*, 2012.

[28] Tobias Becker, Qiwei Jin, Wayne Luk, and Stephen Weston. Dynamic Constant Reconfiguration for Explicit Finite Difference Option Pricing. In *International Conference on Reconfigurable Computing and FPGAs (ReConFig)*, page 176–181, Los Alamitos, CA, USA, Dec. 2011. IEEE Computer Society.

[29] Aric D. Blumer, Henning S. Mortveit, and Cameron D. Patterson. Formal Modeling of Process Migration. In *IEEE International Conference on Field Programmable Logic and Applications (FPL)*, pages 104–110, 2007.

[30] M. Bolić, P. M. Djuric, and S. Hong. Resampling Algorithms for Particle Filters: A Computational Complexity Perspective. *EURASIP*, (15):2267–2277, 2004.

[31] Shekhar Borkar. Designing Reliable Systems from Unreliable Components: The Challenges of Transistor Variability and Degradation. *IEEE MICRO*, pages 10–16, Nov./Dec. 2005.

[32] David Brooks and Margaret Martonosi. Dynamic Thermal Management for High-Performance Microprocessors. In *Proceedings of the International Symposium on High-Performance Computer Architecture*, 2001.

[33] E. Cartwright, A. Fahkari, C. Smith Sen Ma, M. Huang, D. Andrews, and Jason Agron. Automating the Design of MLUT MPSOPC FPGA's in the Cloud. In *International Conference on Field Programmable Logic and Applications (FPL)*, 2006.

[34] Ewerson Carvalho, Ney Calazans, and Fernando Moraes. Heuristics for Dynamic Task Mapping in NoC-based Heterogeneous MPSoCs. *International Workshop on Rapid System Prototyping (RSP)*, 2007.

[35] Charles S. Carver and Michael F. Scheier. Self-focusing Effects of Dispositional Self-consciousness, Mirror Presence, and Audience Presence. *Journal of Personality and Social Psychology*, 36(3):324–332, 1978.

[36] Xun Changqing, Wen Mei, Wu Nan, Zhang Chunyuan, and H.K.-H. So. Extending BORPH for Shared Memory Reconfigurable Computers. In *International Conference on Field Programmable Logic and Applications (FPL)*, 2012.

[37] Serena Chen, Carrie A Langner, and Rodolfo Mendoza-Denton. When Dispositional and Role Power Fit: Implications for Self-expression and Self-other Congruence. *Journal of Personality and Social Psychology*, 96(3):710–727, 2009.

[38] Chen-Yong Cher and Eren Kursun. Exploring the effects of on-chip thermal variation on high-performance multicore architectures. *ACM Trans. Archit. Code Optim.*, 8:2:1–2:22, February 2011.

[39] Eric Cheung, Harry Hsieh, and Felice Balarin. Fast and Accurate Performance Simulation of Embedded Software for MPSoC. In *Proceedings of the Asia and South Pacific Design Automation Conference (ASP-DAC)*. IEEE Press, 2009.

[40] Dan Ciresan, Ueli Meier, and Jürgen Schmidhuber. Multi-column Deep Neural Networks for Image Classification. *IEEE Conference on Computer Vision and Pattern Recognition (CVPR)*, abs/1202.2745, 2012.

[41] Ayse K. Coskun, Tanja S. Rosing, Keith A. Whisnant, and Kenny C. Gross. Static and Dynamic Temperature-Aware Scheduling for Multiprocessor SoCs. *IEEE Transactions on Very Large Scale Integration (VLSI) Systems*, 16(9):1127–1140, 2008.

[42] Matthew Curtis-Maury, James Dzierwa, Christos D. Antonopoulos, and Dimitrios S. Nikolopoulos. Online Power-Performance Adaptation of Multithreaded Programs using Hardware Event-based Prediction. *International Conference on Supercomputing*, 2006.

[43] Marco Dorigo, Vittorio Maniezzo, and Alberto Colorni. Ant System: Optimization by a Colony of Cooperating Agents. *IEEE Transactions on Systems, Man, and Cybernetics, Part B*, 26(1):29–41, 1996.

[44] Arnaud Doucet, Nando de Freitas, and Neil Gordon. *Sequential Monte Carlo Methods in Practice*. Springer, 2001.

[45] Shelley Duval and Robert A Wicklund. A Theory of Objective Self Awareness. *Academic Press*, 1972.

[46] Thomas Ebi, David Kramer, Wolfgang Karl, and Jörg Henkel. Economic Learning for Thermal-aware Power Budgeting in Many-core Architectures. In *Proceedings of the IEEE/ACM/IFIP International Conference on Hardware/Software Codesign and System Synthesis*, CODES+ISSS '11, pages 189–196, New York, NY, USA, 2011. ACM.

[47] Thom J. A. Eguia, Sheldon X.-D. Tan, Ruijing Shen, Eduardo H. Pacheco, and Murli Tirumala. General Behavioral Thermal Modeling and Characterization for Multi-core Microprocessor Design. In *Proceedings of the Conference on Design, Automation and Test in Europe (DATE)*, DATE '10, pages 1136–1141, 3001 Leuven, Belgium, Belgium, 2010. European Design and Automation Association.

[48] T.J. Eguia, S.X.-D. Tan, Ruijing Shen, Duo Li, E.H. Pacheco, M. Tirumala, and Lingli Wang. General Parameterized Thermal Modeling for High-Performance Microprocessor Design. *IEEE Transactions on Very Large Scale Integration (VLSI) Systems*, 20(2):211 –224, Feb. 2012.

[49] Sven Eisenhardt, Thomas Schweizer, Andreas Bernauer, Tommy Kuhn, and Wolfgang Rosenstiel. Prevention of Hot Spot Development on Coarse-Grained Dynamically Reconfigurable Architectures. In *International Conference on Reconfigurable Computing and FPGAs*, 2009.

[50] William J. Froming, G. Rex Walker, and Kevin J. Lopyan. Public and Private Self-awareness: When Personal Attitudes Conflict with Societal Expectations. *Journal of Experimental Social Psychology*, 18(5):476–487, 1982.

[51] Yang Ge, Parth Malani, and Qinru Qiu. Distributed Task Migration for Thermal Management in Many-core Systems. In *Proceedings of the 47th Design Automation Conference*, DAC '10, pages 579–584, New York, NY, USA, 2010. ACM.

[52] Göhringer, D. Meder, L., , M. Hübner, and J. Becker. Adaptive Multi-client Network-on-Chip Memory. In *International Conference on Reconfigurable Computing and FPGAs (ReConFig)*, 2011.

[53] D. Göhringer, M. Chemaou, and M. Hübner. Invited paper: On-chip Monitoring for Adaptive Heterogeneous Multicore Systems. In *7th International Workshop on Reconfigurable Communication-centric Systems-on-Chip (ReCoSoC)*, 2012.

[54] D. Göhringer, M. Hübner, V. Schatz, and J. Becker. Runtime Adaptive Multi-processor System-on-Chip: RAMPSoC. In *IEEE International Symposium onParallel and Distributed Processing (IPDPS)*, 2008.

[55] D. Göhringer, M. Hübner, E.N. Zeutebouo, and J. Becker. CAP-OS: Operating System for Runtime Scheduling, Task Mapping and Resource Management on reconfigurable Multiprocessor Architectures. In *IEEE International Symposium on Parallel Distributed Processing, Workshops and Phd Forum (IPDPSW)*, 2010.

[56] M. Götz, F. Dittmann, and C.E. Pereira. Deterministic Mechanism for Run-time Reconfiguration Activities in an RTOS. In *IEEE International Conference on Industrial Informatics*, 2006.

[57] Marcelo Götz, Florian Dittmann, and Tao Xie. Dynamic Relocation of Hybrid Tasks: Strategies and Methodologies. *Microprocessors and Microsystems*, 33(1):81 – 90, 2009. Selected Papers from ReCoSoC 2007 (Reconfigurable Communication-centric Systems-on-Chip).

[58] Marcelo Götz, Tao Xie, and Florian Dittmann. Dynamic Relocation of Hybrid Tasks: A Complete Design Flow. In *Proceedings of Reconfigurable Communication-centric SoCs*, 2007.

[59] Caroline Goukens, Siegfried Dewitte, and Luk Warlop. Me, Myself, and My Choices: The Influence of Private Self-awareness on Preference-behavior Consistency. *Open Access Publications from Katholieke Universiteit Leuven*, 2007.

[60] Adwait Gupte and Phillip Jones. Hotspot Mitigation using Dynamic Partial Reconfiguration for Improved Performance. In *International Conference on Reconfigurable Computing and FPGAs*, 2009.

[61] F. Hameed, M.A.A. Faruque, and J. Henkel. Dynamic Thermal Management in 3D Multi-core Architecture through Run-time Adaptation. In *Design, Automation Test in Europe Conference Exhibition (DATE), 2011*, pages 1 –6, March 2011.

[62] Hendrik Hangmann. Generating Adjustable Temperature Gradients on Modern FPGAs. *Bachelor's Thesis, University of Paderborn*, Aug 2012.

[63] V. Hanumaiah and S. Vrudhula. Reliability-aware Thermal Management for Hard Real-time Applications on Multi-core Processors. In *Design, Automation Test in Europe Conference Exhibition (DATE), 2011*, pages 1 –6, March 2011.

[64] V. Hanumaiah, S. Vrudhula, and K.S. Chatha. Performance Optimal Online DVFS and Task Migration Techniques for Thermally Constrained Multi-Core Processors. *Computer-Aided Design of Integrated Circuits and Systems, IEEE Transactions on*, 30(11):1677–1690, Nov. 2011.

[65] J. Harden, D. Reese, F. To, D. Linder, C. Borchert, and G. Jones. A Performance Monitor for the MSPARC Multicomputer. In *IEEE Proceedings on SoutheastCon*, 1992.

[66] M. Hatzimihail, M. Psarakis, D. Gizopoulos, and A. Paschalis. A Methodology for Detecting Performance Faults in Microprocessors via Performance Monitoring Hardware. In *IEEE International Test Conference (ITC)*, 2007.

[67] Henry Hoffmann, Jonathan Eastep, Marco D. Santambrogio, Jason E. Miller, and Anant Agarwal. Application Heartbeats: A Generic Interface for Specifying Program Performance and Goals in Autonomous Computing Environments. In *Proceedings of the 7th International Conference on Autonomic Computing*, ICAC '10, pages 79–88. ACM, 2010.

[68] Henry Hoffmann, Jim Holt, George Kurian, Eric Lau, Martina Maggio, Jason E. Miller, Sabrina M. Neuman, Mahmut E. Sinangil, Yildiz Sinangil, Anant Agarwal, Anantha P. Chandrakasan, and Srinivas Devadas. Self-aware Computing in the Angstrom Processor. In *Proceedings of the 49th Annual Design Automation Conference (DAC)*, pages 259–264, 2012.

[69] Chen Huang and Frank Vahid. Dynamic Coprocessor Management for FPGA-enhanced Compute Platforms. *Int. Conf. on Compilers, Architecture and Synthesis for Embedded Systems (CASES)*, 2008.

[70] Kai Huang, Wolfgang Haid, Iuliana Bacivarov, Matthias Keller, and Lothar Thiele. Embedding Formal Performance Analysis into the Design Cycle of MPSoCs for Real-Time Streaming Applications. *ACM Transactions on Embedded Computing Systems*, 11(1), 2012.

[71] Lin Huang and Qiang Xu. Performance Yield-driven Task Allocation and Scheduling for MPSoCs under Process Variation. In *Proceedings of the 47th Design Automation Conference (DAC)*. ACM, 2010.

[72] W. Huang, K. Skadron, S. Gurumurthi, R. J. Ribando, and M. R. Stan. Differentiating the Roles of IR Measurement and Simulation for Power and Temperature-Aware Design. In *Proceedings of the IEEE International Symposium on Performance Analysis of Systems and Software*, 2009.

[73] Wei Huang, Shougata Ghosh, Siva Velusamy, Karthik Sankaranarayanan, Kevin Skadron, and Mircea R. Stan. HotSpot: A Compact Thermal

Modeling Methodology for Early-Stage VLSI Design. *IEEE Transactions on Very Large Scale Integration (VLSI) Systems*, 14(5):501–513, 2006.

[74] IBM. An Architectural Blueprint for Autonomic Computing. Technical report, IBM, 2003.

[75] S. Jovanovic, C. Tanougast, and S. Weber. A Hardware Preemptive Multitasking Mechanism Based on Scan-path Register Structure for FPGA-based Reconfigurable Systems. In *2nd NASA/ESA Conference on Adaptive Hardware and Systems (AHS)*, pages 358–364, 2007.

[76] Heiko Kalte and Mario Porrmann. Context Saving and Restoring for Multitasking in Reconfigurable Systems. In *IEEE International Conference on Field Programmable Logic and Applications (FPL)*, pages 223–228. IEEE, 2005.

[77] Kyungsu Kang, Jungsoo Kim, Sungjoo Yoo, and Chong-Min Kyung. Run-time Power Management of 3-D Multi-Core Architectures Under Peak Power and Temperature Constraints. *Computer-Aided Design of Integrated Circuits and Systems, IEEE Transactions on*, 30(6):905 –918, June 2011.

[78] Ariane Keller, Daniel Borkmann, and Wolfgang Mühlbauer. Efficient Implementation of Dynamic Protocol Stacks. In *Proceeding of the ACM/IEEE Symposium. on Architecture for Networking and Communications Systems (ANCS)*, page 83–84. IEEE Computer Society, 2011.

[79] Ariane Keller, Daniel Borkmann, and Stephan Neuhaus. Hardware Support for Dynamic Protocol Stacks. In *Proceeding of the ACM/IEEE Symposium on Architecture for Networking and Communications Systems (ANCS)*, 2012.

[80] Ariane Keller, Bernhard Plattner, Enno Lübbers, Marco Platzner, and Christian Plessl. Reconfigurable Nodes for Future Networks. In *Proceedings of the IEEE Globecom Workshop on Network of the Future (FutureNet)*, page 372–376. IEEE, 2010.

[81] Jihong Kim and Yongmin Kim. Performance Monitoring and Tuning for a Single-chip Multiprocessor Digital Signal Processor. In *IEEE International Conference on Algorithms and Architectures for Parallel Processing (ICAPP)*, 1996.

[82] Dirk Koch, Christian Haubelt, Thilo Streichert, and Jürgen Teich. Modeling and Synthesis of Hardware-Software Morphing. In *ISCAS*, pages 2746–2749. IEEE, 2007.

[83] Dirk Koch, Christian Haubelt, and Jürgen Teich. Efficient Hardware Checkpointing: Concepts, Overhead Analysis, and Implementation. In *International Symposium on Field Programmable Gate Arrays (FPGA'07)*, pages 188–196, 2007.

[84] Amit Kumar, Li Shang, Li-Shiuan Peh, and Niraj K. Jha. System-Level Dynamic Thermal Management for High-Performance Microprocessors. *IEEE Transactions on Computer-Aided Design of Integrated Circuits and Systems*, 27(1):96–108, 2008.

[85] Rakesh Kumar, Keith I. Farkas, Norman P. Jouppi, Parthasarathy Ranganathan, and Dean M. Tullsen. Single-ISA Heterogeneous Multi-Core Architectures: The Potential for Processor Power Reduction. *International Symposium on Microarchitecture*, 2003.

[86] Eren Kursun and Chen-Yong Cher. Temperature Variation Characterization and Thermal Management of Multicore Architectures. *IEEE MICRO*, pages 116–126, January/February 2009.

[87] H. Kwok-Hay So and R. Brodersen. Runtime Filesystem Support for Reconfigurable FPGA Hardware Processes in BORPH. In *International Symposium on Field-Programmable Custom Computing Machines (FCCM)*, 2008.

[88] Peter R. Lewis, Arjun Chandra, Shaun Parsons, Edward Robinson, Kyrre Glette, Rami Bahsoon, Jim Torresen, and Xin Yao. A Survey of Self-Awareness and Its Application in Computing Systems. In *Proc. Int. Conference on Self-Adaptive and Self-Organizing Systems Workshops (SASOW)*, page 102–107, Ann Arbor, MI, USA, Oct 2011. IEEE Computer Society.

[89] Sergio Lopez-Buedo, Javier Garrido, and Eduardo I. Boemo. Dynamically Inserting, Operating, and Eliminating Thermal Sensors of FPGA-Based Systems. *IEEE Transactions on Components on Packaging Technologies*, 25(4):561–566, 2002.

[90] Enno Lübbers. *Multithreaded Programming and Execution Models for Reconfigurable Hardware*. Logos Verlag, Berlin, 2010.

[91] Enno Lübbers and Marco Platzner. ReconOS: An RTOS Supporting Hard- and Software Threads. In *IEEE International Conference on Field Programmable Logic and Applications (FPL)*. IEEE, Aug 2007.

[92] Enno Lübbers and Marco Platzner. A Portable Abstraction Layer for Hardware Threads. In *IEEE International Conference on Field Programmable Logic and Applications (FPL)*, pages 17–22. IEEE, Aug 2008.

[93] Enno Lübbers and Marco Platzner. Communication and Synchronization in Multithreaded Reconfigurable Computing Systems. In *8th International Conference on Engineering of Reconfigurable Systems and Algorithms (ERSA)*, pages 1–7. CSREA Press, May 2008.

[94] Enno Lübbers and Marco Platzner. Cooperative Multithreading in Dynamically Reconfigurable Systems. In *IEEE International Conference on Field Programmable Logic and Applications (FPL)*, pages 1–4. IEEE, 2009.

[95] Enno Lübbers and Marco Platzner. ReconOS: Multithreaded Programming for Reconfigurable Computers. *ACM Transactions on Embedded Computing Systems (TECS)*, 9(1), October 2009.

[96] Enno Lübbers and Marco Platzner. ReconOS: An Operating System for Dynamically Reconfigurable Hardware. *Dynamically Reconfigurable Systems: Architectures, Design Methods and Applications*, Springer, 2010. ISBN: 978-90-481-3484-7.

[97] Chiao-Ling Lung, Yi-Lun Ho, Ding-Ming Kwai, and Shih-Chieh Chang. Thermal-aware on-line task allocation for 3d multi-core processor throughput optimization. In *Design, Automation Test in Europe Conference Exhibition (DATE), 2011*, pages 1 –6, March 2011.

[98] G. Mariani, V. Sima, G. Palermo, V. Zaccaria, C. Silvano, and K. Bertels. Using Multi-objective Design Space Exploration to Enable Run-time Resource Management for Reconfigurable Architectures. In *Design, Automation Test in Europe Conference Exhibition (DATE)*, 2012.

[99] Dylan McGrath. Analyst: Altera to catch Xilinx in 2012. EE-Times Website: http://www.eetimes.com/electronics-news/4213910/Analyst--Altera-to-catch-Xilinx-in-2012, September 2011.

[100] M. Meterelliyoz, J. P. Kulkarni, and K. Roy. Analysis of SRAM and eDRAM Cache Memories Under Spatial Temperature Variations. *Computer-Aided Design of Integrated Circuits and Systems, IEEE Transactions on*, 29(1):2 –13, Jan. 2010.

[101] Pierre Michaud, André Seznec, Damien Fetis, Yiannakis Sazeides, and Theofanis Constantinou. A Study if Thread Migration in Temperature-Constrained Multicores. *ACM Transactions on Architecture and Code Optimization*, 4(2):1–28, 2007.

[102] J.-Y. Mignolet, V. Nollet, P. Coene, D. Verkest, S. Vernalde, and R. Lauwereins. Infrastructure for Design and Management of Relocatable Tasks in a Heterogeneous Reconfigurable System-On-Chip. pages 986 – 991, 2003.

[103] Alain Morin. Levels of consciousness and Self-awareness: A Comparison and Integration of Various Neurocognitive Views. *Consciousness and Cognition*, 15(2):358–371, 2006.

[104] R. Mukherjee, S. Mondal, and S. O. Memik. Thermal Sensor Allocation and Placement for Reconfigurable Systems. *IEEE/ACM International Conference on Computer-Aided Design*, 4(41):437–442, 2006.

[105] Fabrizio Mulas, David Atienza, Andrea Acquaviva, Salvatore Carta, Luca Benini, and Giovanni DeMicheli. Thermal Balancing Policy for Multi-processor Stream Computing Platforms. *IEEE Transactions on CAD of Integrated Circuits and Systems*, 28(12):1870–1882, 2009.

[106] H. Mushtaq, M. Sabeghi, and K. Bertels. A Runtime Profiler: Toward Virtualization of Polymorphic Computing Platforms. In *International Conference on Reconfigurable Computing and FPGAs (ReConFig)*, 2010.

[107] Vincent Nollet, Prabhat Avasare, Hendrik Eeckhaut, Diederik Verkest, and Henk Corporaal. Run-time Management of a MPSoC Containing FPGA Fabric Tiles. *Transactions on Very Large Scale Integration Systems*, 2008.

[108] Abdullah Nazma Nowroz and Sherief Reda. Thermal and Power Characterization of Field-Programmable Gate Arrays. In *Proceedings of the ACM/SIGDA International Symposium on Field-Programmable Gate Arrays*, 2011.

[109] Abdullah Nazma Nowroz and Sherief Reda. Thermal and Power Characterization of Field-programmable Gate Arrays. In *Proceedings of the 19th ACM/SIGDA international symposium on Field programmable gate arrays*, FPGA '11, pages 111–114, New York, NY, USA, 2011. ACM.

[110] Dongkeun Oh, Nam Sung Kim, Charlie Chung Ping Chen, Azadeh Davoodi, and Yu Hen Hu. Runtime Temperature-based Power Estimation for Optimizing Throughput of Thermal-constrained Multi-core Processors. In *Proceedings of the 2010 Asia and South Pacific Design Automation Conference*, ASPDAC '10, pages 593–599, Piscataway, NJ, USA, 2010. IEEE Press.

[111] M. Oyamada, F.R. Wagner, M. Bonaciu, W. Cesario, and A. Jerraya. Software Performance Estimation in MPSoC Design. In *Asia and South Pacific Design Automation Conference (ASP-DAC)*, 2007.

[112] L.D. Paulson. DARPA Creating Self-aware Computing. *Computer*, 36(3):24, March 2003.

[113] Wesley Peck, Erik Anderson, Jason Agron, Jim Stevens, Fabrice Baijot, and David Andrews. hthreads: A Computational Model for Reconfigurable Devices. In *16th IEEE International Conference on Field Programmable Logic and Applications (FPL)*, pages 885–888. IEEE, Aug 2006.

[114] Rodolfo Pellizzoni and Marco Caccamo. Hybrid Hardware-Software Architecture for Reconfigurable Real-Time Systems. In *Proceedings of IEEE Real-Time and Embedded Technology and Applications Symposium*, 2008.

[115] David Poole, Alan Mackworth, and Randy Goebel. *Computational Intelligence - A Logic Approach*. Oxford University Press, first edition, 1998.

[116] F.J. Rammig, M. Götz, T. Heimfarth, P. Janacik, and S. Oberthur. Real-time Operating Systems for Self-coordinating Embedded Systems. In *9th IEEE International Symposium on Object and Component-Oriented Real-Time Distributed Computing*, 2006.

[117] Jude A. Rivers and Prabhakar Kudva. Reliability Challenges and System Performance at the Architecture Level. *IEEE Design & Test of Computers*, pages 62–72, November/December 200.

[118] Philippe Rochat. Five Levels of Self-awareness as They Unfold Early in Life. *Consciousness and Cognition*, 12(4):717–731, 2003.

[119] Sankalita Saha, Neal K. Bambha, and Shuvra S. Bhattacharyya. A Parameterized Design Framework for Hardware Implementation of Particle Filters. *IEEE Int. Conf. on Acoustics, Speech and Signal Processing*, 2008.

[120] V. Salapura, K. Ganesan, A. Gara, M. Gschwind, J.C. Sexton, and R.E. Walkup. Next-Generation Performance Counters: Towards Monitoring Over Thousand Concurrent Events. In *IEEE International Symposium on Performance Analysis of Systems and software (ISPASS)*, 2008.

[121] Aswin C. Sankaranarayanan, Rama Chellappa, and Ankur Srivastava. Algorithmic and Architectural Design Methodology for Particle Filters in Hardware. *Int. Conf. on Computer Design*, 2005.

[122] Marco D. Santambrogio, Henry Hoffmann, Jonathan Eastep, and Anant Agarwal. Enabling Technologies for Self-aware Adaptive Systems. In *NASA/ESA Conference on Adaptive Hardware and Systems (AHS)*, pages 149–156. IEEE, 2010.

[123] M.A. Sayed and P.H. Jones. Characterizing Non-ideal Impacts of Reconfigurable Hardware Workloads on Ring Oscillator-Based Thermometers. In *International Conference on Reconfigurable Computing and FPGAs (ReConFig)*, pages 92 –98, Dec. 2011.

[124] Hartmut Schmeck, Christian Müller-Schloer, Emre Cakar, Moez Mnif, and Urban Richter. *Organic Computing — A Paradigm Shift for Complex Systems Autonomic Systems*, chapter Adaptivity and Self-organisation in Organic Computing Systems, pages 5–37. Volume 1 of [16], 2011.

[125] G. Di Marzo Serugendo, M.-P. Gleizes, and A. Karageorgos. *Self-Organizing Software – From Natural to Artificial Adaptation*. Springer, 2011.

[126] Seyab and Said Hamdioui. NBTI Modeling in the Framework of Temperature Variation. In *Proceedings of the Conference on Design, Automation and Test in Europe*, DATE '10, pages 283–286, 3001 Leuven, Belgium, Belgium, 2010. European Design and Automation Association.

[127] S. Sharifi and T.S. Rosing. Accurate Direct and Indirect On-Chip Temperature Sensing for Efficient Dynamic Thermal Management. *Computer-Aided Design of Integrated Circuits and Systems, IEEE Transactions on*, 29(10):1586 –1599, Oct. 2010.

[128] Kamana Sigdel, Mark Thompson, Andy D. Pimentel, Carlo Galuzzi, and Koen Bertels. System-level Runtime Mapping Exploration of Reconfigurable Architectures. *Proceedings of the International Symposium on Parallel & Distributed Processing (IPDPS)*, 2009.

[129] V.-M. Sima and K. Bertels. Runtime Decision of Hardware or Software Execution on a Heterogeneous Reconfigurable Platform. In *IEEE International Symposium onParallel Distributed Processing (IPDPS)*, 2009.

[130] V.-M. Sima, E.M. Panainte, and K. Bertels. Resource Allocation Algorithm and OpenMP Extensions for Parallel Execution on a Heterogeneous Reconfigurable Platform. In *International Conference on Field Programmable Logic and Applications (FPL)*, 2008.

[131] H. Simmler, L. Levinson, and R. Männer. Multitasking on FPGA Coprocessors. In *IEEE International Conference on Field Programmable Logic and Applications (FPL)*, pages 121–130. Springer-Verlag, 2000.

[132] F. Sironi, D. B. Bartolini, S. Campanoni, F. Cancare, H. Hoffmann, D. Sciuto, and M. D. Santambrogio. Metronome: Operating System Level Performance Management via Self-adaptive Computing. In *Proceedings of the 49th Annual Design Automation Conference (DAC)*, pages 856–865. ACM, 2012.

[133] F. Sironi, A. Cuoccio, H. Hoffmann, M. Maggio, and M.D. Santambrogio. Evolvable Systems on Reconfigurable Architecture via Self-aware Adaptive Applications. In *NASA/ESA Conference on Adaptive Hardware and Systems (AHS)*, pages 176 –183, 2011.

[134] F. Sironi, M. Triverio, H. Hoffmann, M. Maggio, and M.D. Santambrogio. Self-aware Adaptation in FPGA-based Systems. *International Conference on Field Programmable Logic and Applications (FPL)*, 2010.

[135] K. Skadron, M.R. Stan, W. Huang, Sivakumar Velusamy, Karthik Sankaranarayanan, and D. Tarjan. Temperature-aware Microarchitecture. In *Proceedings of the 30th Annual International Symposium on Computer Architecture*, 2003.

[136] Lodewijk T. Smit, Johann L. Hurink, and Gerard J. M. Smit. Run-time Mapping of Applications to a Heterogeneous SoC. *Proceedings of the International Symposium on System-on-Chip*, 2005.

[137] Lodewijk T. Smit, Gerard J. M. Smit, Johann L. Hurink, Hajo Broersma, Daniel Paulusma, and Pascal T. Wolkotte. Run-time Assignment of Tasks to a Heterogeneous Processors. *Embedded Systems Symposium*, 2004.

[138] H.K.-H. So, A. Tkachenko, and R. Brodersen. A Unified Hardware/Software Runtime Environment for FPGA-based Reconfigurable Computers using BORPH. In *International Conference on Hardware/Software Codesign and System Synthesis (CODES+ISSS)*, 2006.

[139] Hoyden Kwok-Hay So and R.W. Brodersen. Improving Usability of FPGA-Based Reconfigurable Computers Through Operating System Support. In *International Conference on Field Programmable Logic and Applications (FPL)*, 2006.

[140] R. Sterritt and M. Hinchey. SPAACE IV: Self-Properties for an Autonomous and Autonomic Computing Environment - Part IV A Newish Hope. In *IEEE International Conference and Workshops on Engineering of Autonomic and Autonomous Systems (EASe)*, 2010.

[141] Greg Stitt, Roman Lysecky, and Frank Vahid. Dynamic Hardware/Software Partitioning: A First Approach. *Proceedings of the Design Automation Conference (DAC)*, 2003.

[142] Andrew S. Tanenbaum. *Modern Operating Systems*. Prentice-Hall, Inc., second edition, 2001.

[143] G. A. Tawney. Feeling and Self-awareness. *Psychological Review*, 9(6):570–596, 1902.

[144] S. Traboulsi, Wenlong Zhang, D. Szczesny, A. Showk, and A. Bilgic. An Energy-efficient Hardware Accelerator for Robust Header Compression in LTE-Advanced Terminals. In *International Conference on Field Programmable Logic and Applications (FPL)*, 2012.

[145] S. Uhrig, S. Maier, G. K. Kuzmanov, and T. Ungerer. Coupling of a Reconfigurable Architecture and a Multithreaded Processor Core with Integrated Real-time Scheduling. In *International Parallel and Distributed Processing Symposium (IPDPS)*, 2006.

[146] Osman S. Unsal, James W. Tschanz, Keith Bowman, Vivek De, Xavier Vera, Antonio González, and Oguz Ergin. Impact of Parameter Variations on Circuits and Microarchitecture. *IEEE MICRO*, pages 30–39, November/December 2006.

[147] S. Vassiliadis, S. Wong, G. Gaydadjiev, K. Bertels, G. Kuzmanov, and E.M. Panainte. The MOLEN Polymorphic Processor. *IEEE Transactions on Computers*, 53(11):1363 – 1375, 2004.

[148] Siva Velusamy, Wei Huang, John Lach, Micrea Stan, and Kevin Skadron. Monitoring Temperature in FPGA based SoCs. In *Proceedings of the IEEE Int. Conf. on Computer Design*, 2005.

[149] Hai Wang, Sheldon X.-D. Tan, Guangdeng Liao, Rafael Quintanilla, and Ashish Gupta. Full-chip Runtime Error-tolerant Thermal Estimation and Prediction for Practical Thermal Management. In *Proceedings of the International Conference on Computer-Aided Design*, ICCAD '11, pages 716–723, Piscataway, NJ, USA, 2011. IEEE Press.

[150] Wenping Wang, Shengqi Yang, S. Bhardwaj, S. Vrudhula, F. Liu, and Yu Cao. The Impact of NBTI Effect on Combinational Circuit: Modeling, Simulation, and Analysis. *IEEE Transactions on Very Large Scale Integration (VLSI) Systems*, 18(2):173 –183, Feb. 2010.

[151] Michael J. Wooldridge. *An Introduction to MultiAgent Systems*. Wiley, second edition, 2009.

[152] Maury Wright. Roving Reporter: Processors and FPGAs Match Well in Data-intensive Applications. `http://embedded.communities.intel.com/community/en/hardware/blog/2011/03/07/roving-reporter-processors-and-fpgas-match-well-in-data-intensive-applications`, March 2011.

[153] Yen-Kuan Wu, S. Sharifi, and T.S. Rosing. Distributed Thermal Management for Embedded Heterogeneous MPSoCs with Dedicated Hardware Accelerators. In *Computer Design (ICCD), 2011 IEEE 29th International Conference on*, pages 183 –189, Oct. 2011.

[154] Xilinx. Virtex-5 Family Overview, Data Sheet DS100 (v5.0). `http://www.xilinx.com/support/documentation/data_sheets/ds100.pdf`, February 2009.

[155] Xilinx. Spartan-6 Family Overview, Data Sheet DS160 (v2.0). `http://www.xilinx.com/support/documentation/data_sheets/ds160.pdf`, October 2010.

[156] Xilinx. Virtex-4 Family Overview, Data Sheet DS112 (v3.1). `http://www.xilinx.com/support/documentation/data_sheets/ds112.pdf`, August 2010.

[157] Xilinx. Virtex-5 FPGA Data Sheet: DC and Switching Characteristic, Data Sheet DS202 (v5.3). `http://www.xilinx.com/support/documentation/data_sheets/ds202.pdf`, May 2010.

[158] Xilinx. Virtex-5 FPGA System Monitor, User Guide UG192 (v1.7.1). `http://www.xilinx.com/support/documentation/user_guides/ug192.pdf`, February 2011.

[159] Xilinx. 7 Series FPGAs Overview, Data Sheet DS180 (v1.12). `http://www.xilinx.com/support/documentation/data_sheets/ds180_7Series_Overview.pdf`, October 2012.

[160] Xilinx. Embedded System Tools Reference Manual: EDK, User Guide UG111 (v14.3). `http://www.xilinx.com/support/documentation/sw_manuals/xilinx14_3/est_rm.pdf`, October 2012.

[161] Xilinx. MicroBlaze Processor Reference Guide: Embedded Development Kit EDK 14.3, User Guide UG081 (v14.3). `http://www.xilinx.com/cgi-bin/SW_Docs_Redirect/sw_docs_redirect?locale=en&ver=14.3&topic=sw+manuals&sub=mb_ref_guide.pdf`, October 2012.

[162] Xilinx. Virtex-6 Family Overview, Data Sheet DS150 (v2.4). `http://www.xilinx.com/support/documentation/data_sheets/ds150.pdf`, January 2012.

[163] Xilinx. Xilinx Using EDK to Run Xilkernel on a MicroBlaze Processor, User Guide UG758 (v14.1). `http://www.xilinx.com/support/documentation/sw_manuals/xilinx14_1/ug758.pdf`, April 2012.

[164] Inchoon Yeo and Eun J. Kim. Temperature-Aware Scheduler Based on Thermal Behavior Grouping in Multicore Systems. In *Proceedings Design, Automation, and Test in Europe*, 2009.

[165] Inchoon Yeo, Chih C. Liu, and Eun J. Kim. Predictive Dynamic Thermal Management for Multicore Systems. In *Design Automation Conference*, 2008.

[166] Chunbo Zhang, Ramachandra Kallam, Andrew Deceuster, Aravind Dasu, and Leijun Li. A Thermal-mechanical Coupled Finite Element Model with Experimental Temperature Verification for Vertically Stacked FPGAs. *Microelectronic Engineering*, 91:24–32, March 2012.

[167] Yufu Zhang and A. Srivastava. Accurate temperature estimation using noisy thermal sensors for gaussian and non-gaussian cases. *Very Large Scale Integration (VLSI) Systems, IEEE Transactions on*, 19(9):1617 –1626, Sept. 2011.

[168] Yufu Zhang, Ankur Srivastava, and Mohamed Zahran. On-chip Sensor-driven Efficient Thermal Profile Estimation Algorithms. *ACM Transactions on Design Automation of Electronic Systems*, 15:25:1–25:27, June 2010.

[169] Kenneth M. Zick and John P. Hayes. On-Line Sensing for Healthier FPGA Systems. In *Proceedings of the ACM/SIGDA International Symposium on Field-Programmable Gate Arrays*, 2010.

[170] Kenneth M. Zick and John P. Hayes. Low-cost Sensing with Ring Oscillator Arrays for Healthier Reconfigurable Systems. *ACM Transaction Reconfigurable Technology and Systems*, 5(1):1:1–1:26, Mar. 2012.